Taunton's

PLUMBING
COMPLETE

EXPERT ADVICE FROM START TO FINISH

The Taunton Press, Inc.,
63 South Main Street, PO Box 5506,
Newtown, CT 06470-5506
e-mail: tp@taunton.com

Editor: Martin Miller
Copy Editor: Candace B. Levy
Indexer: Jay Kreider
Jacket/Cover design: Kimberly Adis
Interior design: Kimberly Adis
Layout: Cathy Cassidy
Illustrator: Trevor Johnston
Photographer: Rex Cauldwell, except where noted

Library of Congress Cataloging-in-Publication Data
Cauldwell, Rex.
  Plumbing complete / Rex Cauldwell.
      p. cm.
  Includes index.
  ISBN 978-1-56158-855-8
  1. Plumbing. 2. Dwellings--Maintenance and repair. I. Title.
  TH6123.C38 2009
  696'.1--dc22
                              2008053007

Printed in the United States of America
10 9 8 7 6 5 4 3 2 1

To my daughter, Melissa, and my two grandchildren, Katy and Elizabeth

ACKNOWLEDGMENTS

It has been said hundreds of times in as many books—that one book is not the work of just one person—it is a collaboration of many persons with many talents. My thanks goes all those who worked on the book at The Taunton Press book division and to those outside that organization who work with them—Martin Miller, for example, who actually put my ramblings in a readable form. Thanks too, to all the suppliers who provided material for me to build with and destroy—and provided photographs of their products: Kohler®, Danco®, American Standard®, Cash Acme®, all the good folks at Rocky Mount Supply of Rocky Mount VA., Vanguard®, Ridge Tools, Studor®, InSinkErator®, Sioux Chief Manufacturing, Watts®, Zoeller® Pump Company, and Oatey® Products. And last but never least, my gratitude goes to Kimberly Lavender, the tradesperson whose hands and body grace some of the pages of this book.

—Rex Cauldwell

contents

CLEANING AND UNCLOGGING 72

REPAIRING FAUCETS 94

>>   >>   >>   >>

TOOLS

Y OU PROBABLY ALREADY HAVE many of the plumbing tools you'll need—right out in your garage. A hacksaw, for example, is a good general-purpose tool for cutting most anything. So is a utility knife. And don't forget flashlights, levels, a miter box, screwdrivers, prybars, putty knife, wire brushes, extension cords, caulking guns, hammers, files, rulers, sledge hammer, cold chisels, dolly, ladders, chisels, sawhorses, and other tools in your toolbox.

While many do-it-yourself jobs will require only a few simple hand tools, others require—and become a lot easier—with specialized tools. There's no need to get overwhelmed, however. As you expand your skills, you will also expand your plumber's toolkit. Remember, if a job requires a tool you've never heard of or have no idea how to operate, call a professional or do a little online or library research before tackling the job.

HAND TOOLS

POWER TOOLS

CLEANING TOOLS

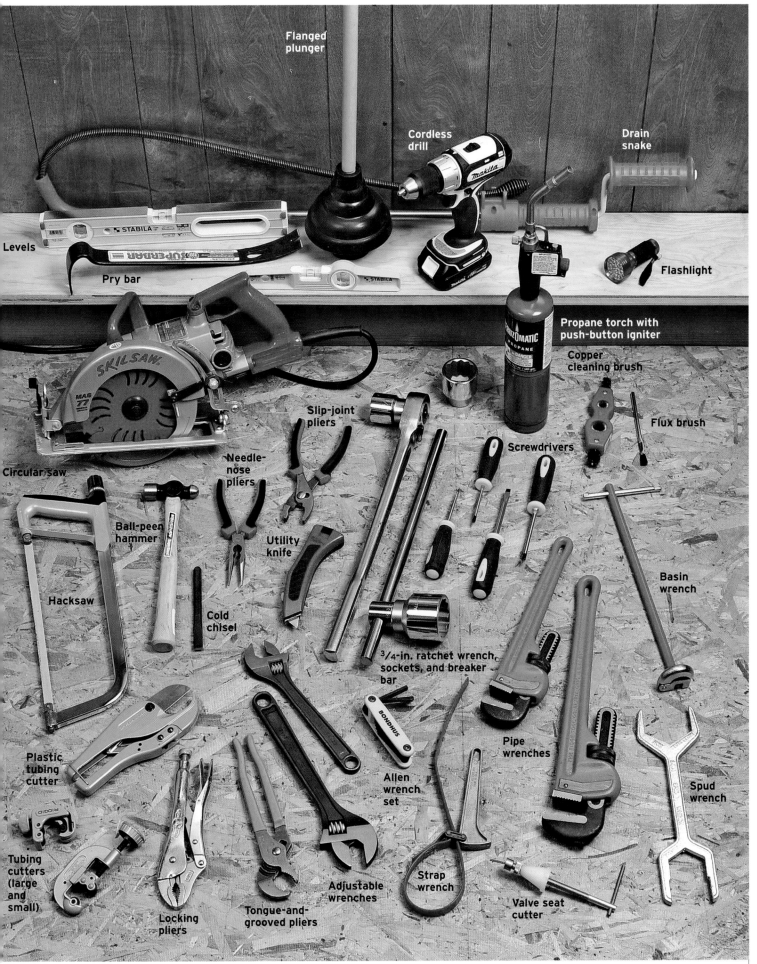

Flanged plunger

Cordless drill

Drain snake

Levels

Pry bar

Flashlight

Propane torch with push-button igniter

Copper cleaning brush

Flux brush

Slip-joint pliers

Screwdrivers

Circular saw

Needle-nose pliers

Ball-peen hammer

Utility knife

Basin wrench

Hacksaw

Cold chisel

³/4-in. ratchet wrench, sockets, and breaker bar

Plastic tubing cutter

Pipe wrenches

Allen wrench set

Spud wrench

Tubing cutters (large and small)

Locking pliers

Tongue-and-grooved pliers

Adjustable wrenches

Strap wrench

Valve seat cutter

WRENCHES AND PLIERS

Pipe wrenches are still the mainstay of the plumbing trade–with their angled teeth, they can grab round objects like pipe, nipples, and galvanized couplings. They come in 2-in. increments from 6 in. to 18 in. and up. The more you have, the better, but at the very least you need one small (6 in.), one medium (12 in.), and one large (18 in.). Even better, get two of each size. Modern pipe wrenches come in aluminum and with angled heads.

The larger the wrench, the more leverage and the easier the work. The common steel pipe wrench has a jaw 90 degrees to the handle. However, special pipe wrenches perform specific plumbing tasks more effectively than a 90-degree wrench.

These jaws are spring loaded and won't easily slip off a pipe.

Its steel jaws and an aluminum handle make this wrench light (but more expensive). (Photo courtesy of Ridge Tool Company)

With jaws offset about 45 degrees, this pipe wrench can be turned in tight locations. (Photo courtesy of Ridge Tool Company)

This specialty wrench has its handle set in line with the jaws. (Photo courtesy of Ridge Tool Company)

Always use a back-up wrench– with each engaged in opposite directions–when turning fittings on any kind of threaded pipe.

Adjustable wrenches

Adjustable wrenches turn flat-sided objects. Available in three sizes—small, medium, and large—they can replace a whole collection of box- or open-end wrenches. To avoid rounding a nut, always tighten the jaws firmly.

Pliers

Pliers grip things, too—pipes, nuts, rods, and faucet parts. They also turn, bend, and press objects together. Pliers fall into three categories: fixed pivot (like needle-nose pliers), movable pivot (like tongue-and-grooved pliers), and those that lock on an object when you apply increased pressure (locking pliers). You'll need at least one slip-joint pliers, one needle-nose model, a couple of tongue-and-grooved pliers, and medium-sized locking pliers for most plumbing jobs.

Hex, Torx®, and star wrenches

Some wrenches come in specific profiles for loosening screws, cap screws, and some socket nuts. You'll use hex, Torx, and star wrenches to remove faucet handles and for other repairs. They come L-shaped or with T-handles and also in folding sets. Get a folding set of each style—the housing keeps everything in one place and functions as the handle.

Socket wrenches

Socket wrenches encase the entire circumference of a bolt or plumbing fixture and allow more torque to be applied without stripping or rounding the faces. They're especially useful for removing rusted elements, anode rods, and other tasks for which extra muscle is needed to break a fastener loose. You might get by with a few selected sizes (such as $1\frac{1}{2}$ in. and $1\frac{1}{16}$ in.), but a full $\frac{3}{4}$-in. set and a breaker bar will keep you from running to the hardware store. Sockets and wrenches are expensive. Look for them at yard sales and auctions to keep the costs down.

Adjustable wrenches with wide handles are easier on your hands than those with narrow handles.

Some common pliers you'll want in your plumbing toolbox. From left to right: slip-joint pliers, medium-size tongue-and-grooved pliers, large tongue-and-grooved pliers, needle-nose pliers, and locking pliers.

Hex (or Allen) wrenches, along with Torx (also called "star") wrenches, are an increasing necessity in the modern toolbox.

A $\frac{3}{4}$-in. socket set and breaker bar will remove the most stubborn nuts and elements.

SPECIALTY TOOLS

Some plumbing situations call for tools developed specifically for those jobs. For example, old faucet handles have a nasty habit of corroding on the stem, making them hard, if not impossible, to remove. Here's where a handle puller comes to the rescue. And once inside an older faucet, you may want to remove the seat or just grind it smooth with a valve-seat grinding tool. You'll find speciality tools, such as a valve-seat wrench, lock-nut wrench, faucet-handle puller, seat cutter, spud wrench, strainer wrench, and pipe vise, worth the investment. Sometimes you just can't get the job done with any other tool.

➜ See "Faucet Repair Tools," p. 97.

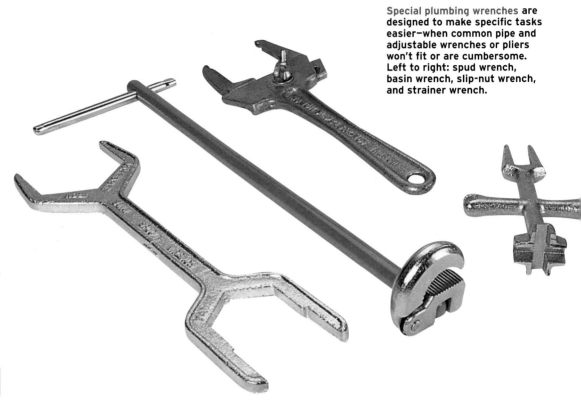

Special plumbing wrenches **are designed to make specific tasks easier—when common pipe and adjustable wrenches or pliers won't fit or are cumbersome. Left to right: spud wrench, basin wrench, slip-nut wrench, and strainer wrench.**

Using a basin wrench

Both pipe wrenches and adjustable wrenches need a lot of horizontal room, and sometimes that room does not exist—for example, when you're under a sink attaching or removing a supply tube to a faucet. At best, you may have 4 in. in which to maneuver a tool. Enter the basin wrench. It extends vertically and has jaws on top set at 90 degrees to the handle. The perpendicular jaws allow this wrench to be used in tight locations.

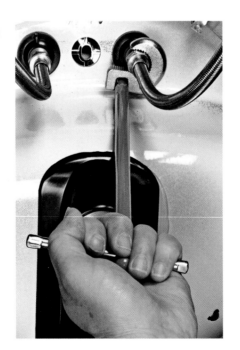

Moen's cartridge puller can remove **cartridges you can't get out of the faucet valve body with ordinary grab-and-pull pliers.**

For many plumbing jobs, **especially those involving galvanized pipe, you'll need a third hand—something to anchor the pipe securely while you're applying torque with two hands on a pipe wrench. These situations call for the firm grip of a pipe vise.**

Valve seat tools

If you suspect the cause of a leak in an old-style faucet is a bad seal at the seat, you can get the seat out with a valve seat wrench (left). Rough seats will destroy a new seal, making the valve leak rather quickly. And if a little smoothing will put it back in business, a seat grinder (top center) will do the job.

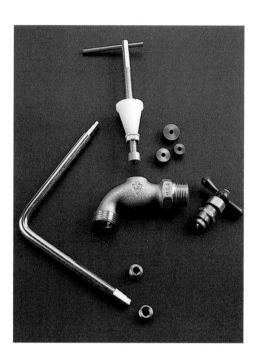

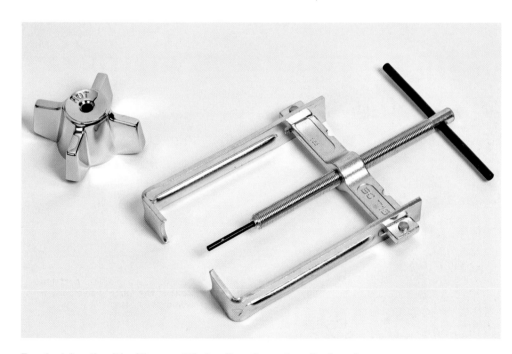

For stuck handles, **it's either use this handle puller or toss the faucet.**

A strap wrench uses **a fiber strap that locks around an object. It works well on unusually shaped items, such as tub spouts and hard-to-remove glue-can caps.**

COPPER AND PLASTIC PIPE TOOLS

Because different types of pipe require different installation methods, specific tools have been developed for each material. Sometimes, one tool will work on more than one kind of pipe, but in all cases you'll produce more professional results by using the right tool, instead of just the one that's handy.

Copper pipe tools

Sweating copper requires a multitude of tools. Besides a pipe or tubing cutter, you will need plumber's cloth, paste, solder, and a flame shield. These specialized tools, if used properly, will produce strong, long-lasting joints that don't leak.

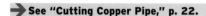

➡ **See "Cutting Copper Pipe," p. 22.**

➡ **See "Sweating Copper Pipe," p. 24.**

Tools for plastic pipe

Plastic pipe comes in several forms—extruded polyethylene (PEX), polyethylene, chlorinated polyvinyl chloride (CPVC), and polyvinyl chloride (PVC). Many tools, like ratcheting scissors, can be used on more than one kind of pipe material, but others, like crimpers, are designed for one type of pipe only. In a pinch, you can cut some plastics with a sharp pocket or utility knife, but a fine-tooth hacksaw or chopsaw blade intended for plastics will also do the trick. In addition to tools for cutting and crimping, you'll need glues for PVC and CPVC pipe.

Flaring tools

Small tubing cutter

Tubing cutter

Propane torch with push-button igniter

Copper cleaning brush multi-purpose tool

Plumber's cloth

Flux brush

Soldering flux

Spring tubing benders

Solder

Fiberglass flame shield

When choosing tools for cutting plastic pipe, **metal ratcheting scissors (left) are preferred over non-ratcheting scissors (right).**

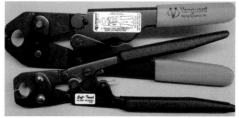

These crimpers from Vanguard **come with short handles but they close with the leverage of long handles.**

PEX crimpers typically come with short handles (bottom two crimpers) for tight locations, but they are hard to close. Long-handled models (top) close easily but can't be used in cramped quarters.

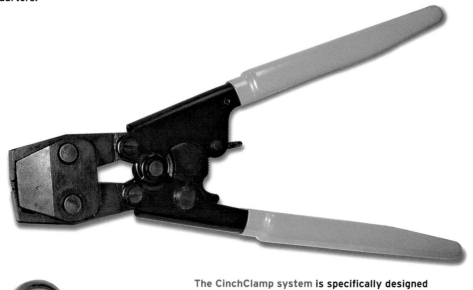

The CinchClamp system is specifically designed for crimping stainless-steel rings around PEX pipe and onto PEX fittings. It produces solid leakproof joints with a precision copper banding doesn't offer. Look for them on the internet. (Photo courtesy of Watts Radiant, Inc.)

The CinchClamp stainless ring comes with a rectangular section that fits the jaws of the crimper precisely. This leaves the body of the crimp ring completely visible, allowing you to position it precisely before crimping it to the PEX pipe. (Photos courtesy of Watts Radiant, Inc.)

DRILLS AND SAWS

A cordless drill is an indispensable tool for all DIY home-improvement jobs. Once you have one, you won't know how you got along without it. A quick-change sleeve will let you go back and forth between drilling and fastening tasks without constantly stopping to open the chuck.

You'll need a high-quality circular saw when cutting studs and for many other rough-in plumbing chores. Blades come made for a wide variety of cutting tasks, from ripping to fine cutting. Diamond sawblades are available and are great for cutting cast-iron and steel pipe.

Jigsaw blades are made in a number of configurations and from different metals for different tasks. Make sure your blade matches the job you want it to do.

A hammer drill combines rotary and hammering actions to make quick work out of light-duty masonry drilling. You'll need a corded or cordless model when drilling holes in block walls for masonry fasteners.

For both new work and remodeling, you'll need a collection of power tools. Cutting; fastening; and drilling studs, joists, and blocking will, at a minimum, call for a circular saw and cordless drill. Masonry work will demand a rotary hammer or hammer drill. If you have only periodic need for an expensive tool, consider renting it.

Cordless tools

Cordless tools are great for working in a crawlspace and on roofs and ladders. Many have completely replaced corded tools on the jobsite simply because they're so convenient. Higher-voltage tools produce great torque but their large batteries can make them heavy. Smaller voltage tools are lighter but don't have the kick of their more-powerful brothers.

AC corded tools

There are times when you need the torque of a corded tool. The most common are the motorized miter saw (chopsaw), hammer drill, circular saw, reciprocating saw, jigsaw, and right-angle drill (medium and heavy duty). All these need to be powered with a short ground-fault circuit interrupter (GFCI) extension cord. Corded tools have the extra torque needed when cutting with diamond sawblades on steel and cast-iron pipes and when drilling with diamond hole cutters in tile. Diamond-tipped hole cutters (available online) are essential for drilling tile; carbide-tipped bits won't work.

This rotary hammer is the tool you'll need for cutting large holes in concrete.

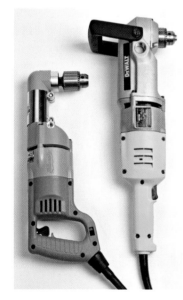

Pistol-grip right-angle drills are light and easy to use, especially when working overhead, but their torque is limited. Heavy-duty right-angled drills put a lot of power into the drill bit, but if the bit jams in the wood, that power can jerk your hand and arm against studs or framing with enough force to break a bone.

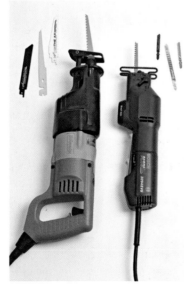

A full sized reciprocating saw makes cutting pipe and demolition work easy, but a mini-reciprocating saw is better for small cuts in metal because of its extremely thin blade. It's also good for working in tight locations.

All types of drill bits come in various qualities. Always buy the best bits you can afford. Those with a higher price tag will prove a worthwhile investment; they'll last a long time.

UNCLOGGING TOOLS

The tool you choose to unclog a plumbing fixture will depend on the nature of the clog and the fixture itself. In many cases, you can unblock a stopped sink, toilet, or other fixture with a good-quality plunger. More stubborn clogs will send you for one of several kinds of augers. If roots in the exterior drain line are the culprit, you can rent a rodding machine, but it's often easier just to call in the pros.

➜ **See "Cleaning and Unclogging," p. 72.**

A hand-turned auger is operated by a revolving handle and is useful for clearing sinks and small-diameter pipes. (Photo courtesy of Ridge Tool Company)

A pistol-grip cordless auger works just like a manual model, but a battery-powered motor does the work. (Photo courtesy of Ridge Tool Company)

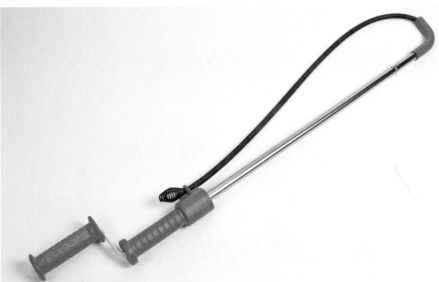

A toilet auger, with its long coiled line, can often unclog a blocked toilet when a plunger won't do the job. Toilet augers come in 3-ft. and 6-ft. lengths.

Electric augers are equipped with a cutterhead made to slice through tree roots. Its long line will reach exterior blockages. (Photo courtesy of Ridge Tool Company)

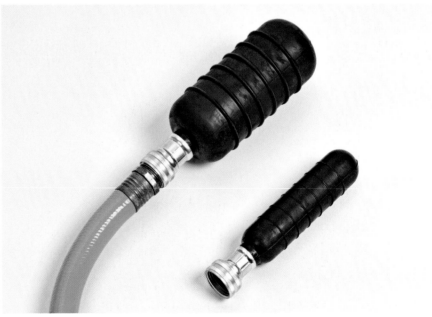

Drain-cleaning bladders come in two sizes, the smaller diameter for drains up to 1½ in. and the larger one for drains in 2 in. to 4 in. lines. The bladder attaches to a hose, and when you turn the water on, it expands, sealing the drain and sending a high-pressure jet of water out the front, which blows out the stoppage.

— PICKING A PLUNGER —

A trip to the hardware store can net you a half dozen different plunger designs—from throwaways to high-quality tools. A good plunger must make a perfect seal in the bowl of the fixture you're trying to unclog. When you push down and pull up, the end needs to stay in full contact with the surface of the bowl. In addition, the plunger has to be large enough to move a lot of water.

These plungers have a soft end **that makes a good seal, but they can only pull and push a small amount of water.**

The flange on this plunger folds out, **adapting it to different drains. Its large bulb makes a good seal and is able to move a significant amount of water.**

Plungers with pleated ends **like this will move a massive amount of water, but their thin, hard plastic will not seal tightly against the surface of a drain.**

Safety first
Just as important as getting a plumbing job done is getting it done safely. A sloppy plumber can encounter plenty of accidents—not to mention hazardous materials. Never scrimp on tools—especially those that will keep you safe. Use the right tools; vent gases properly; don't be too proud to call a plumber; and keep a list of utility, fire, police, and medical emergency numbers nearby.

GFCI-protected extension cord

Fire extinguisher

Safety goggles

Leather gloves

Heavy-duty gloves

Be sure to wear safety glasses, **especially when working overhead. The goggles will keep debris from falling into your eyes.**

SUPPLY PIPES

INSTALLING OR REPAIRING THE PIPES that carry water throughout a house requires skills well within the reach of the average homeowner. With today's modern labor-saving materials, one can plumb the water-supply system in a house in a single day. As this chapter will show you, you don't have to solder if you want copper pipe, and you don't have to glue or crimp if you want plastic pipe. Today's push-on fittings have made all that old hat. Before you start on even a moderate repair job, however, it's wise to look at the path your water pipes take from their entry into your home to the various fixtures. That will give you a mental map of your water-supply system and will put the job you're doing in a practical perspective.

PLANNING AND PREPARATION

COPPER PIPE

PLASTIC PIPE

THREADED PIPE

FITTINGS AND VALVES

REPAIR AND INSTALLATION

THE WATER-SUPPLY SYSTEM

Water enters your house from a municipal or public water system or, in many rural areas, from a well. Whatever its source, water flows through lines categorized as either "service" pipe, which gets the water to the house, or "distribution" pipe, which takes the water from the main entry point to the various fixtures (sinks, dishwashers, toilets, water heaters, etc.).

Pipe sizes

The larger the pipe, the more water it can carry, so in most cases, service pipe is 1 in. or 1¼ in. in diameter. This diameter ensures that all parts of the house will have a sufficient supply, even when you're running the dishwasher in the kitchen and someone's showering in the upstairs bath.

Distribution pipes are typically 1 in., ³⁄₄ in., or ½ in. in diameter. The larger diameters of distribution pipe are usually used to carry water from room to room. The smaller diameters are generally found bringing water to various fixtures, like sinks.

Pipe material

In the old days, service pipe was either galvanized pipe or copper. Neither are used much today due to their cost and tendency to corrode. Most of today's service pipe is either polyethylene or PVC. Typically, for average size residential installations, polyethylene service pipe is used in up to 1-in. diameters. PVC in larger diameters is used for large residential and commercial jobs. The advantage of polyethylene is that it comes in long rolls so you can install the entire run from a well or municipal meter base to the house without splices. There's nothing wrong with using PVC as a service pipe, however. It just takes longer to install, requires a splice every 10 ft. to 12 ft., and cracks easily if laid on a rock.

Modern distribution pipe, like service pipe, also has evolved. Galvanized is out and copper is slipping fast, giving way to its nonmetal counterparts—CPVC and PEX. When you're looking for distribution pipe in a house that has undergone previous repairs, you may see a mix of different materials, depending on when the work was done.

Tracing your water flow

The first element that water going into your home encounters in most municipal systems is the water meter. It may be located inside the house or outside in what's sometimes called a "buffalo box," and It may have valves on one or both sides. It may also have electrical connections hooked up to a self-reporting system, which tells your water company how much you've used each month.

Once inside the house, the line branches off to the water heater and from there, both hot and cold water travel in pairs to the various fixtures in your house. Along the way, you may find the cold water line branching off to outdoor water faucets, and both hot and cold water going to upper stories through vertical pipes called "risers."

You may also notice that the distribution pipe changes diameter from time to time at "reducing fittings," depending on the fixture it's supplying.

Along the lines and at the fixtures, you will also notice different valves, such as stop valves, ball valves, and saddle Ts, which control the flow of the water. Some fixtures, however, may not feature stop valves, which shut off water to individual faucets or fixtures, and installing these valves should be one of your first priorities in doing any plumbing renovation or repair.

Most water meters are at or close to the entry point of the water system into the house. Look under a metal, concrete, or plastic cover.

Water heater lines must be full-sized metallic lines—typically ³⁄₄ in. Reducing these will lower your volume and pressure.

Both hot and cold water are fed to the upper stories of a house through vertical pipes called risers.

TYPICAL WATER-SUPPLY AND DISTRIBUTION SYSTEM

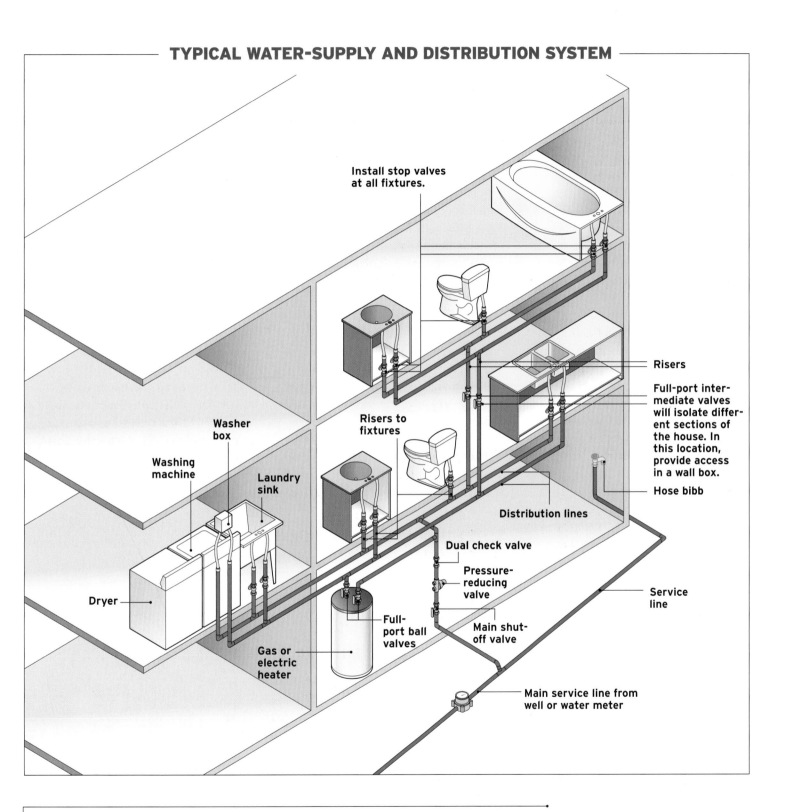

Install stop valves at all fixtures.

Washer box

Washing machine

Laundry sink

Risers to fixtures

Dryer

Full-port ball valves

Gas or electric heater

Dual check valve

Pressure-reducing valve

Main shut-off valve

Distribution lines

Risers

Full-port intermediate valves will isolate different sections of the house. In this location, provide access in a wall box.

Hose bibb

Service line

Main service line from well or water meter

TRADE SECRET

When storing any pipe, be sure to plug the ends to keep out insects. Store CPVC and PEX pipe out of the sunlight—UV rays will gradually deteriorate plastics.

SHUTTING OFF THE WATER

If there's any mantra you should adopt before working on a water line it's "shut off the water." Not doing so can leave you and your surrounding workspace both soaked and unsafe. Just exactly where you shut the water off depends on the kind of fixture you're working on, the kind of pipe that supplies it, and the presence or absence of shut-off valves.

Working backward

If you're working on a fixture, such as a sink or toilet, look under the fixture for shut-off valves ("stop" valves) on both the cold and the hot water lines. Most homes built in the last 4 decades or so have them—generally with chrome-plated oval handles or round handles with indented "knuckles."

Stop valves are made for light, infrequent duty, and if they don't completely stop the water, don't try to tighten them with a pliers. They will likely break, and you'll have defeated your original purpose. If there's no stop valves on a fixture, include their installation on your to-do list, then look back down the system for intermediate valves.

Intermediate valves

First, disregard any valves on pipes coming from a hot water heating system boiler. These aren't the cause of any water-supply problems. Then follow the path of water lines in the basement or crawlspace. Look for valves with a round or straight handle and turn the appropriate one(s) off (hot, cold, or both). If the water has been used recently, you can tell which carries the hot water by feeling it. If both pipes feel cold, trace them back to the water heater to make sure.

Other spots to look for intermediate valves are in tub and shower access panels, behind the bathtub itself, and on the water heater. Access panels can be tricky to find. Look in the back of the wall in which the pipes are located (often in a closet). Watch for ridges along a painted joint or round marks indicating screws buried in the paint.

One of the problems intermediate valves pose is that you can't be immediately sure what part of the house they control. To find out, close one at a time and turn on the faucets in the house section by section. Flush the toilets, too—twice. (A toilet will flush once with the water in its tank.) Then, pull the tank lid off and look to see if it's filling.

If you can't find intermediate valves, you'll have to shut off the water at the main valve.

The main valve

All homes have at least one main whole-house shut-off valve. Sometimes (especially in freezing climates) the valve is inside the home, generally close to the water meter.

If you don't have a water meter, look for a large pipe that enters near the basement floor level and follow it to the valve. It you still can't find a valve, then look outside.

Outdoors, a main shut-off valve is located in a box (a stop box or buffalo box) on a line that branches off the municipal line. In areas with sidewalks, these boxes are often located in the grassy area between the sidewalk and the street. You may have to poke along the surface to find the box, but its lid, if not visible, will usually be only a few inches below the surface.

Although you can get the lid off with a prybar or adjustable wrench, in freezing climates, the actual valve will be located below the frost line. To operate the valve in these cases, you'll have to get a "key" (a special wrench head on a long steel shaft) from your utility or hardware store.

Once you've located the valve that shuts the water off to the fixture or section of pipe you're working on, you need to drain the affected system of any residual water.

➡ See "Repairing and Tapping Water Lines," p. 44.

Stop valves installed at the fixture **allow you to turn off the water at the fixture itself, rather than to the whole house.**

A main shut-off valve **is often located on one or both sides of the water meter and shuts the water off to the whole house.**

Intermediate valves **control the flow of water to the house on a section-by-section basis.**

FINDING LEAKS

The bane of plumbing is a water leak–the quiet drip that doesn't make noise. It falls silently onto the floor or down a wall, and its damage can be extensive. Finding it means observing potential leaky spots.

Look to the floor on the trim and carpet **A**. Once you find puddled water or wet carpet, look behind the wall and trim for the culprit. Then go "up" to find the leak **B**. Once you've located the potential source of the leak, examine the pipe and fittings. Water flows downhill, so don't assume the first wet fitting you find is the leaky one. Follow the pipe upward to the topmost wet fitting **C**. To verify a slow leak is coming from a high fitting, dry off the fitting and place a piece of paper towel behind it **D**. To verify a leak at a low fitting, dry off the pipe above and the fitting itself. If the pipe above is dry and a drop of water forms on the fitting, that fitting is the problem **E**.

Shut off the water and drain the lines. Disassemble the fitting and analyze the cause. Look for an out-of-round pipe, burrs on a cut edge, unsquare cuts, or improperly tightened fittings **F**. Fix the problem and reinstall the fitting, using a new one if necessary.

A Look to the floor for evidence of slow leaks in pipes and fittings.

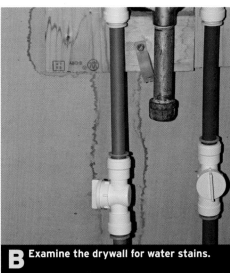

B Examine the drywall for water stains.

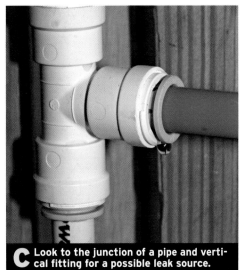

C Look to the junction of a pipe and vertical fitting for a possible leak source.

D A paper towel behind a vertical fitting can test for water leakage.

E Wipe the fitting dry and watch for the reappearance of water at a junction.

F A leak at any galvanized fitting means you must replace that section of pipe.

COPPER PIPE AND FITTINGS

Copper is an all-metal pipe that comes in two basic types: rigid (also called "hard" or "stick" copper) or rolled (also called "soft") pipe. For new residential plumbing, most plumbers use rigid copper. For remodel plumbing, with a lot of existing framing in the way, you might consider soft copper, but only if local plumbing codes won't allow CPVC or PEX pipe. Soft copper will kink very easily as you unroll or bend it to fit a corner. The biggest use of soft copper is in HVAC systems.

➤ See "Bending Copper Pipe," p. 23.

➤ See "PEX and Pipe Fittings," p. 31.

Copper grades

Copper pipe comes in different thicknesses for different uses. Over time, copper dissolves from the inside out, and as it does, it creates pinholes. The thickest copper will last longest before it springs a leak. Copper thickness grades are indicated by different colored lettering on the pipe. Red lettering designates the pipe as type M, the thinnest and most commonly found grade in residential usage. Blue lettering is designated as type L (indicating a medium grade or thickness) and is more expensive. Type K pipe, the thickest and most expensive, carries green lettering and is not used in residential plumbing.

Copper sweat fittings

Fittings orient a pipe in the direction you want it to go, and for copper pipe, they come in two types: sweat fittings (both in copper and brass) and push-on fittings. Sweat fittings attach to pipe using the old tried-and-true methods of solder, flux, and flame. Push-on fittings grab the exterior of the pipe and eliminate the time and effort sweating requires.

➤ See "Push-On Fittings," p. 40.

WARNING

The smaller the pipe diameter, the thinner its wall. Thus the tiny copper pipe used for the refrigerator ice maker has a wall not much thicker than a piece of paper and will easily develop pinholes from corrosion in active water (water with a low pH)—sometimes within 6 months. Use plastic pipe instead.

When buying copper, **you will face the choice of stick lengths (left) or rolls. Rolled copper will come in boxes (right).**

The color of the lettering on copper pipe **indicates its grade, the thickness of its wall.**

Boiler drain

Ball valve

Ball valve with side drain

Gate valve

The most common types of copper sweat valves. **The threaded boiler drain (top left) also accepts copper pipe for sweating.**

Copper fittings come full sized or as reducers, used when you need to change the diameter of supply pipe

Some sweat fittings can be made only in brass, such as fittings with drains, street fittings, in-line threaded taps, and drop-ear elbows.

T-fittings divert the flow of water at right angles to its flow and are used to run water from one line to another or to a fixture.

Both 90-degree- and 45-degree-angle fittings change the in-line direction of the water flow.

For sweating copper together with copper, use common couplings or unions (upper right in photo).

SIZING WATER LINES—RULES OF THUMB

- Make the main line from municipal supply to the house 1 in. Don't reduce to $3/4$ until after the main turnoff valve which must be a full-flow ball valve.

- Make all lines in the house $3/4$ in. except those running to a single fixture which can be $1/2$ in.

- For each additional full bath, run $3/4$-in. line from the main cutoff ball valve and control it with a full-size ball valve. For hot water at the additional full bath, run a separate $3/4$-in. line for it from the water heater.

- If a second full bath is at a remote location from the first full bath, consider adding a separate water heater at its location to shorten the run of the hot water line. You can do the same for the kitchen at a remote location.

Local requirements may vary. Always check your local codes.

CUTTING COPPER PIPE

You can cut copper with a hacksaw, a mini-reciprocating saw with a fine-tooth metal-cutting blade, or with a tubing cutter— whose opposing wheels gradually shear the pipe as you turn the tool around it. A tubing cutter produces clean cuts, square to the pipe.

Place the tubing cutter on the pipe between the cutter's two rollers, and turn the knob on the handle to snug the rollers against the pipe. Do not tighten excessively. ❶ Turn the tubing cutter around the pipe several times, making sure the wheels stay in a single track with each revolution. Once the cutter feels loose, turn the knob to resnug the cutter wheels. Then turn the cutter around the pipe several more times. Follow this procedure until the pipe snaps ❷. The cutting wheel leaves a sharp edge inside the pipe. Flatten it with the reamer that comes on the cutter ❸. If you've misjudged the length of a pipe you've cut and need to trim it, a tubing cutter won't work; it will slip off the end of the pipe. Instead, use a chopsaw or a mini-reciprocating saw with a fine-tooth metal-cutting blade.

➡ See "Trimming a pipe end," p. 29.

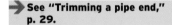

⚠ **WARNING**

Nails and screws can easily pierce copper or plastic pipe, so you have to install the pipes beyond the reach of the fasteners. Run pipes in the center of studs, 1¼ in. from the front face of the framing. If not possible, install protective metal plates on the stud face.

1 Place the cutter on the pipe and turn the handle to snug it on the pipe.

2 Turn the cutter around pipe, resnugging it as you go. Repeat until the pipe is cut.

3 Use the reamer on the tubing cutter to remove the ridge the cutter raises on the inside edge of the pipe end.

ALTERNATE REAMING TECHNIQUES

When a tubing cutter doesn't have a reaming tool or the reamer is worn, use a rat-tail file (below left) or a stepped drill bit (below right) to clean the cut.

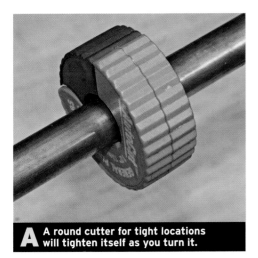

A A round cutter for tight locations will tighten itself as you turn it.

B Tighten the knob on the cutter as you turn it.

C Shim the pipe slightly away from the wall and use a fine-tooth blade to cut it.

Cutting copper in tight spaces

When a tight space won't give you enough room to turn a tubing cutter, use a round cutter **A** or a midget cutter **B**. A round cutter will tighten itself as you turn it. To tighten the midget cutter, use a pliers to turn the small knob.

When a copper pipe is flush against a wall and has to be cut, not even cutters made for tight locations will work. To solve this problem, use a wood shim to keep the pipe away from the wall about 1/8 in. Cut the pipe with a fine-toothed blade. Though a hacksaw will work, a mini-reciprocating saw makes the job infinitely easier **C**.

BENDING COPPER PIPE

A Using spring benders to bend rolled copper can prevent kinking the pipe.

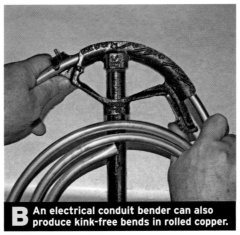

B An electrical conduit bender can also produce kink-free bends in rolled copper.

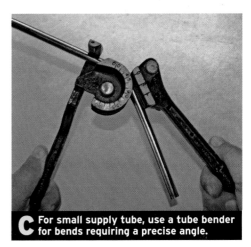

C For small supply tube, use a tube bender for bends requiring a precise angle.

Rolled copper pipe needs a form or tool to create a bend without kinks. Bending by hand normally won't work. Typically, spring benders, which slip over the pipe, are used **A**. An electrical conduit bender also works well as a form for the bend **B**. Use a bender one size larger than the pipe diameter: a 1/2-in. bender for 3/8-in. pipe or a 3/4-in. bender for 1/2-in. stock. For very small diameter copper pipe, such as a supply tube for a sink or toilet, use a mechanical bender made for this purpose **C**. These benders are marked in degrees and will produce very accurate bends.

SWEATING COPPER PIPE

The most common problem you'll experience after sweating copper pipe is a leaking connection due to an improperly cleaned pipe or fitting. To prevent this, clean both with either a wire brush made for cleaning copper or a plumber's cloth (sandpaper on a fabric backing). Clean the outside of the pipe slightly beyond the point where the fitting's hub will terminate. ❶ A new fitting is not clean. It has an oil film that must be removed with a wire brush made to fit into its hub ❷. Before you heat the pipe and fitting and apply solder, brush a soldering paste on both the pipe and the fitting ❸. The paste allows the solder to flow properly. A tinning paste leaves a metal powder in the joint and allows the solder to flow even better. Once you've applied paste to both the fitting and the pipe, slide the fitting on the pipe. Wipe away any excess paste—not doing so may cause the solder to flow away from the fitting ❹. Now you are ready to heat the pipe and fitting. First, look at the work area. Can anything be damaged by fire or heat? If fire, you need to insert a fiberglass antiburn cloth to block the flames. If heat is going to be a problem, you will have to use a 1-in. concrete block behind the flame. ❺ Apply heat first to the fitting and then to the pipe, alternating back and forth until both are hot enough to melt the solder. If you get the joint too hot, you will ruin the paste, and the solder will not flow. If the joint is not hot enough, nothing will happen. It takes some practice to determine how much heat you should apply in a given time ❻. Place something to catch any solder drips under the joint. Keep tapping the solder onto the joint until the solder starts to melt. Once the solder melts, let around 1/2 in. of solder flow into joint and remove any excess solder ❼.

1 Using a wire brush, clean the outside of the copper pipe.

2 Clean the inside hub of a fitting also, even if the fitting is new.

5 Insert an antiflame cloth between the pipe and any surface that flame will damage. Temporarily nail it in place, if necessary.

TRADE SECRETS

• Use a low-temperature solder with a wide working range (such as Oatey Safe-Flo) and a hard-to-burn tinning paste (such as Oatey #95) to aid soldering.

• If water in a pipe has frozen, it may have changed the pipe's outside diameter, and fittings will not slide on. You will have to cut the pipe back to where it has not frozen to get a perfectly round circumference for fittings. Sometimes however, you can use a tubing flaring tool to squeeze the pipe back to proper diameter for a short distance.

• Solder will flow vertically upwards into a joint.

3 Apply solder paste to the outside of the pipe and to the inside hub of the fitting.

4 Wipe away any excess paste from the fitting and pipe.

6 Apply heat to the fitting hub and the pipe. Move the flame back and forth to heat the joint evenly.

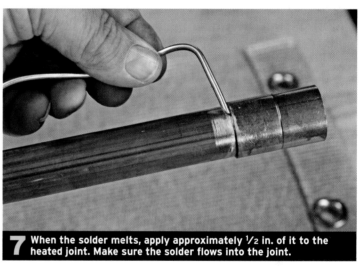

7 When the solder melts, apply approximately ½ in. of it to the heated joint. Make sure the solder flows into the joint.

COPPER CLEANERS

When cleaning copper, use plumber's cloth (pliable fabric-backed sandpaper on a roll) or wire brushes designed to clean both the outside of a pipe and the inside of a fitting.

Get waterproof plumber's cloth. It will not dissolve when wet, or contaminate the pipe or fitting. Wire brushes come individually or as an all-in-one cleaning tool. Keep both plumber's cloth (open weave or solid) and wire brushes handy.

⚠ WARNING

A fiberglass flame shield or piece of metal stops only the flame, it does not stop the heat from igniting flammable materials. To stop heat penetration, use a 1-in.-thick concrete block.

CPVC PIPE AND FITTINGS

Chlorinated polyvinyl chloride (CPVC), a semi-flexible pipe for indoor use, is commonly referred to as "plastic" pipe, because it was the first non-metallic pipe to challenge copper as a pipe material. CPVC installs quickly and easily with CPVC glue or push-on fittings. It won't dissolve like copper, but it will split faster than copper when subjected to freezing temperatures. It's not affected by acidic water (water with a low pH, which will corrode metals).

CPVC male and female adapters

Because CPVC is a brittle plastic, adaptors with all-plastic threads exhibit some weaknesses when paired with metal fittings. Male and female adaptors tend to leak at the threads, and both female and male threads tend to crack at the shoulder when tightened. To avoid this, always insert a piece of pipe in the fitting as it is screwed into a fixture. To completely alleviate such problems, avoid all-plastic threaded fittings. Instead, use metal-threaded CPVC male and female adapters.

Turning corners

As with other fittings, you can turn a corner at either 45 degrees or 90 degrees with CPVC fittings. When installing a fitting on pipe, use a fitting with the same size hubs on both ends. When connecting to another CPVC fitting, use a street elbow or street 45-degree fitting. These fittings have a hub on one end sized for pipe insertion and a smaller-diameter hub on the other end, sized to slip into another fitting hub.

CPVC valves

There are three types of CPVC in-line valves. The round-handled valve is a globe valve and should be avoided because it slows down the water flow and doesn't conform to many local plumbing codes. A ball valve is the preferred valve. However, the metal valves designated for CPVC tend to leak around the handle, and the all-plastic models tend to jam or turn off very hard. The best CPVC valves are female-threaded ball valves with two CPVC brass male adapters and Sharkbite-type push on valves.

Couplings, caps, and reducers

You can connect CPVC pipe together with glued couplings. You can even glue PVC to CPVC using a special white coupling. If you need a reducer coupling or don't have one, you can glue a reducer bushing on a full-sized coupling to create

CPVC, a plastic pipe, **comes in the same sizes as other supply pipe and installs quickly and easily.**

CPVC male **and female adapters. CPVC transition fittings with brass threads help avoid split collars common when using all-plastic threads.**

a reducer coupling. You can even put a reducer bushing or coupling on each side of a 3/4-in. coupling to create a 1/2-in. coupling.

CPVC drop-ear elbows

CPVC drop-ear elbows are used to terminate a CPVC pipe and start a threaded pipe. Use these for the through-the-wall pipe to the shower head and any time you want to come through a wall with a finished pipe to a stop valve under the sink or toilet.

>> >> >>

Insert pipe in CPVC fitting before tightening.

Never tighten a male or female adapter without a pipe inserted into its hub. This prevents ovaling and cracking of the hub.

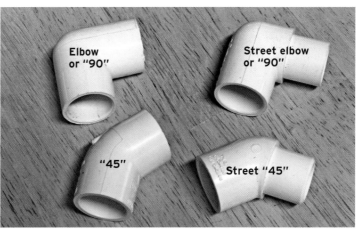

Elbow or "90"

Street elbow or "90"

"45"

Street "45"

CPVC angles. Fittings with the same size hubs are used in-line with pipe. Those with one smaller hub are used to connect to another fitting.

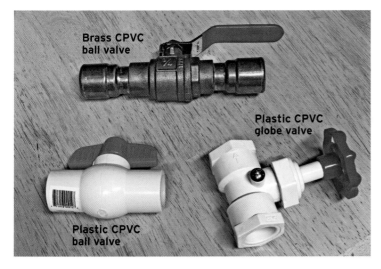

Brass CPVC ball valve

Plastic CPVC globe valve

Plastic CPVC ball valve

CPVC valves. The plastic valves can jam or require extra force when turning them off. The brass valve won't jam and provides CPVC fittings.

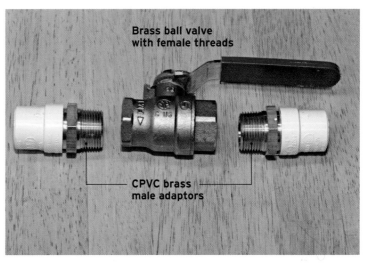

Brass ball valve with female threads

CPVC brass male adaptors

Two CPVC brass male adapters plus a threaded ball valve make a high-quality, CPVC-compatible valve.

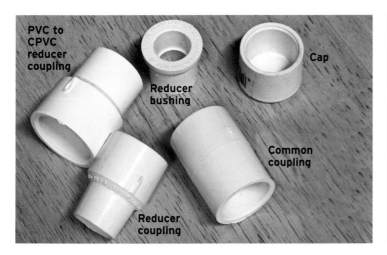

PVC to CPVC reducer coupling

Reducer bushing

Cap

Common coupling

Reducer coupling

CPVC couplings are made to join sections of same-size pipe. Reducer couplings transition from one pipe size to another.

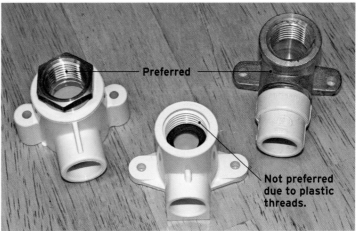

Preferred

Not preferred due to plastic threads.

Use CPVC drop-ear elbows when you need to terminate a length of CPVC pipe and start a threaded pipe of different material.

CPVC PIPE AND FITTINGS (CONTINUED)

CPVC T-fittings

CPVC T-fittings are used to send water off to the sides of the main line, from a ³/₄-in. main line to a ½-in. line for the toilet, for example, or at the end of a ³/₄-in. line from which you want to branch one or two ½-in. lines to various fixtures.

CPVC unions

A CPVC union is used wherever you want a removable coupling—for example, at an appliance that might need to be replaced, like a water heater. It is very easy, however, to avoid CPVC unions (and any potential leaking plastic threads) by simply cutting a pipe and gluing a coupling back into the line.

Interfacing

You can interface CPVC with PEX with a special fitting that glues onto CPVC and has PEX barbs on the opposite end. You can interface CPVC with copper or PEX using push-on fittings. You can also interface CPVC or PEX with threaded galvanized fittings and nipples using male and female adapters.

➡ See "Push-On Fittings," p. 40.

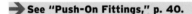

Plastic pipe tips

- **Do not subject CPVC pipe to extreme vibration in the winter (for example, transporting it tied to the top of a truck). Vibration and shock may crack the pipe.**
- **CPVC gets brittle with age. To cut old pipe (and keep it from cracking), you may need to use a fine-tooth sawblade instead of ratcheting scissors (which compress the pipe).**
- **Use a black-diamond blade sharpener to keep the ratcheting scissors blade razor sharp.**
- **Do not plumb CPVC directly to a water heater on either the hot or cold supply lines. At this juncture, use a 12-in. to 18-in. by ³/₄-in. flexible stainless-steel braided connector.**

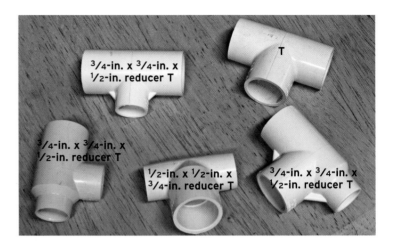

CPVC T-fittings are configured to accept pipe of the same size or as reducer fittings in installations requiring branch lines of smaller sizes.

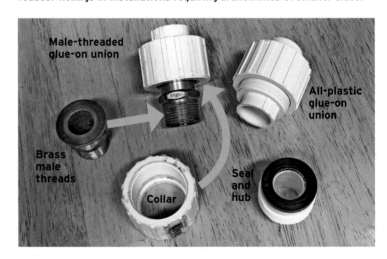

CPVC unions allow you to disconnect two lengths of pipe at their junction.

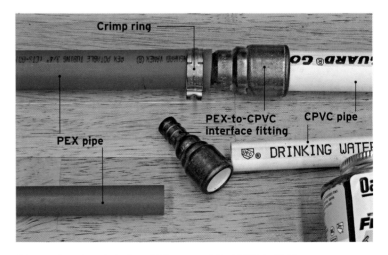

Glue-on-to-crimp interface fittings can join CPVC to PEX.

CUTTING CPVC PIPE

CPVC pipe cuts easily, but if you use a hacksaw—even with a fine-tooth blade—you'll need to smooth the burrs from the cut edge. A solid-blade chopsaw works great, but you have to bring the pipe to it; and this cut too, will need smoothing. For all CPVC cutting, a ratcheting scissors is the most convenient tool.

The first rule when using ratcheting scissors is to keep the blade razor sharp. A dull blade will compress the pipe and crack it. To keep blades sharp, use a diamond whetstone (it has a black color) ❶. Though you can cut new CPVC by simply by closing the scissors handle, older CPVC pipe can be brittle, requiring you to arc the cut to keep from cracking the pipe. The following technique will avoid cracking in most cases. First, close the CPVC scissors just far enough to slightly cut into the pipe and cause the pipe to compress or oval slightly. Do not close the handles all the way unless you are sure the pipe will cut and not crack. ❷ As the blade starts to cut into the pipe, push the cutter in a downward direction while you turn the pipe upward in an clockwise motion ❸. This splits the pipe, instead of cracking it.

After CPVC has been installed or stored for a while, it cracks easily when cut with any type of scissors blade. In this case, use a very fine tooth blade in a mini-reciprocating saw.

1 Keep the blade of the CPVC pipe cutter razor sharp with a diamond whetstone.

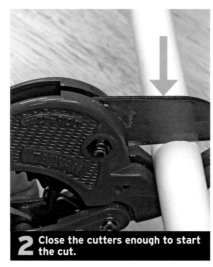

2 Close the cutters enough to start the cut.

3 Push the cutter downward and turn the pipe as you cut.

Trimming a pipe end

Cutting a pipe in the middle is one thing, but trimming off ¹/₈ in. is another thing entirely. A tubing cutter won't work on copper; it will slip off. A ratcheting scissors can't cut CPVC that close either. Only two things will work. A solid-blade chopsaw can trim ¹/₁₆ in. off both copper and CPVC. A very fine tooth blade can trim a tiny sliver off either pipe, if needed.

GLUING CPVC PIPE

Your first task when gluing CPVC is to pick a glue. You have two choices—orange glue or gold glue. There is essentially no difference in the performance of either type, but the gold glue will save you a little time. It doesn't require a primer. The orange glue does.

If you're going with the gold, apply the gold glue both inside the fitting hub and on the outer surface of the pipe end ❶. Though the entire surface of the hub and pipe end needs to be coated, don't over-apply the glue. Typically, a pro will go around both the fitting and the pipe twice and, on the third time, will wipe away the excess with the dauber. Immediately after you apply the glue, slip the fitting onto the pipe, or vice versa. A pro will slip the fitting on slightly out of alignment and then turn either the fitting or the pipe into proper alignment ❷. This spreads the glue evenly around the joint. Don't rush, but don't waste any time, either. These glues set up quickly.

If you're using glue with primer, apply the primer (it's purple) with essentially the same application techniques to coat the surfaces. The primer dulls the plastic and gives the glue a "tooth" for better adhesion. Then apply the glue and assemble the joint.

Gold glue (left) does not require a primer. Orange glue (right) does, so its application is a two-step process. Use in a ventilated area. Glue vapors can be harmful.

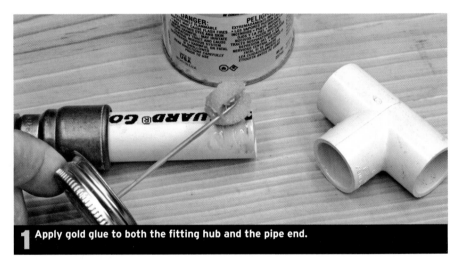

1 Apply gold glue to both the fitting hub and the pipe end.

⚠ WHAT CAN GO WRONG

Too much CPVC glue may form a dam inside the pipe or fitting once they're assembled. To avoid the excess, rake the dauber off in the can, leaving just enough glue on it to put a thin coat on both the fitting and the pipe.

2 Assemble the fitting and pipe fully, slightly out of orientation. Turn the pieces to the proper orientation to spread the glue. Hold for several seconds.

PEX PIPE AND FITTINGS

Extruded polyethylene (PEX) pipe has taken the place of polybutylene (PB) pipe, which does not meet current codes because of its tendency to spring leaks. PEX is flexible and can take a moderate freeze without splitting. Different manufacturers make PEX for specialized purposes, such as in-floor radiant heating and for outdoor use, so be sure you're purchasing a product designed for your specific needs. PEX installs quickly and easily with specialized fittings and crimp bands and even more easily with push-on fittings.

➡ See "Push-On Fittings," p. 40.

Like CPVC, PEX is not affected by acidic water. However, you should not use it to plumb directly to a water heater, either on the cold or the hot water outlet sides. Plastic and metal have different expansion rates, and such a connection will leak. For hot water heater installations, use a 12-in. to 18-in. length of 3/4-in. flexible stainless-steel braided connector to interface the two.

PEX fittings

PEX fittings have barbed ends and come in brass and plastic. The plastic versions are made to resist chlorine and acid water. Brass fittings won't crack, but consider the plastic fittings for areas that have acidic water, which will corrode copper.

Fittings are available for almost any situation, including water-line taps to refrigerators and various types of valves. However, PEX fittings allow only 90-degree corners—there are no PEX 45-degree fittings. If one is required, you can make it by inserting two male adapters in a galvanized or brass 45-degree fitting. Small brass splitters that change the direction of the water flow are available when you need extra taps beyond what a common T-fitting will provide.

>> >> >>

PEX is a flexible plastic **pipe that installs very quickly and easily. It is a do-it-yourselfer's dream.**

Ts and elbows. **Use these fittings when you want to tap off the main line or turn a corner.**

PEX fittings (left to right): **a PEX-to-copper sweat interface, PEX-to-female thread (for showerheads and stop valves), and a plug to keep debris out of the pipe until it's used.**

PEX PIPE AND FITTINGS (CONTINUED)

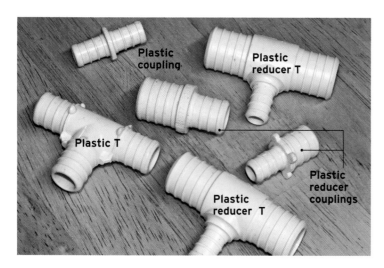

Plastic coupling

Plastic reducer T

Plastic T

Plastic reducer T

Plastic reducer couplings

Plastic PEX fittings. Use these in acidic water where the water corrodes copper pipe.

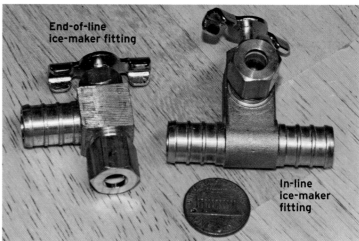

End-of-line ice-maker fitting

In-line ice-maker fitting

PEX ice-maker fittings. Use these and non-metallic line instead of thin-walled soft copper tubing to connect your ice maker to the water supply.

PEX tips

- Use push-on fittings to speed installation.
- Keep PEX at least 6 in. away from all heat sources, such as hot-water-heater vents.
- Do not plumb directly to a water heater.
- Do not make severe turns without an elbow; sharp turns will kink PEX.
- Install crimped fittings per manufacturer's instructions. Most leaks come from improperly crimped fittings.
- Crimper must be held at 90 degrees to the pipe.
- Crimper jaws must cover all of the ring as it is crimped.

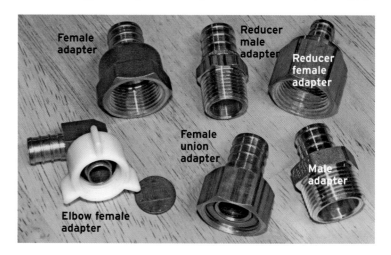

Female adapter

Reducer male adapter

Reducer female adapter

Female union adapter

Male adapter

Elbow female adapter

PEX male and female adapters. Use these to transition PEX pipe to pipe of other materials.

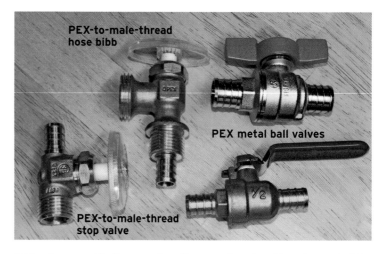

PEX-to-male-thread hose bibb

PEX metal ball valves

PEX-to-male-thread stop valve

PEX valves provide the same function as valves for other pipe material with the ease of PEX installation. Sharkbite-type valves can also be used.

PEX splitters provide outlets for more than the single tap a common T-fitting provides.

PEX crimp rings

PEX pipe is commonly installed using black copper crimp bands. The black color differentiates the PEX bands from copper-colored PB bands, which are no longer allowed by national plumbing codes. To crimp the bands, you will need a crimping tool, either a long- or a short-handled model. Long-handled crimpers provide good leverage, which makes crimping easy. Short-handled crimpers can be used in tight locations where the others won't fit, but the handles on most models are hard to close.

➡ See information on crimping tools, pp. 8-9.

Some plumbers prefer stainless-steel crimp rings over the copper bands because they generally install more quickly, and crimping the stainless bands requires less effort. For these, you'll have to get a special crimper with a proprietary head made to fit the stainless rings. In addition, stainless rings may prove difficult to find. Look for them on the Internet or at a large local plumbing-supply outlet.

➡ See "Crimping Stainless Bands," p. 35.

CREATING A 45

A 45-degree PEX fitting can be created by using a 45-degree brass fitting and two PEX male adapters. Wrap the threaded ends with Teflon® tape and screw them into the 45.

PEX crimp rings. Both black copper rings (left), and stainless-steel crimp rings (right) require specialized crimping tools.

CRIMPING PEX

PEX cuts easily with a sharp ratcheting scissors. Position the scissors square to the pipe and push the handles together to cut through the pipe ❶. Insert the copper ring over the pipe ❷. Don't worry about placement of the ring at this time. Slide the fitting into the pipe as far as the fitting will allow. Some fittings have physical stops built in, which makes this task foolproof. Always install the rings with the procedures recommended by the manufacturer of the PEX pipe you're using. Most PEX varieties require you to slide the ring up against the fitting shoulder, then back it off the shoulder by about ¼ in. However, some manufacturers design their fittings so the ring bumps up against the shoulder and stays there ❸. Using a pliers, gently squeeze the ring tight against the pipe ❹. At this stage, the ring must be tight enough to keep the crimpers from moving it out of place as you begin to crimp it. The most common error is crimping the ring out of place. After verifying the ring is still positioned properly, place the crimper jaws over the ring, making sure it lies at 90 degrees to the pipe ❺. Close the jaws of the crimper completely. You will feel the ring compress. The crimper must not move the ring out of place. Open the jaws of crimpers all the way. Visually check the ring to make sure it's the proper distance from the fitting shoulder and that it's been crimped fully around its circumference and not at an angle ❻. If the ring is imperfectly crimped, you'll have to cut the pipe, remove the ring and start over.

1 Using a sharp ratcheting scissors, cut through the PEX pipe.

2 Insert the crimp ring over the pipe.

3 Slide the fitting into the pipe and orient the fitting in the proper direction.

4 Using a pliers, squeeze the ring gently to keep it in place.

5 Using the crimping tool, crimp the ring onto the pipe without changing the position of the ring.

6 Verify that the crimp has been made properly and that the ring is in the proper place.

CRIMPING STAINLESS BANDS

Slide the stainless band onto the pipe **❶**. Insert the fitting into the PEX pipe **❷**. Adjust the stainless band so it's ¼ in. from the fitting shoulder (or at the distance recommended by the manufacturer) and crimp the band with the crimping tool **❸**. Make sure the band is placed correctly on the fitting, square to the pipe and crimped properly around its entire circumference **❹**.

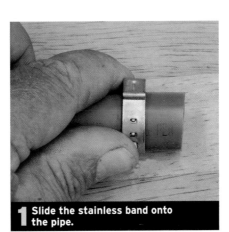

1 Slide the stainless band onto the pipe.

2 Insert the fitting into the pipe.

3 Adjust the band, ¼ in. from the fitting shoulder and crimp the band.

4 Make sure that the band is positioned and crimped properly.

THREADED PIPE AND FITTINGS

Male threads are found on the outside of the fitting, while threads on the inside are called female.

Pipe nipples come in ½ in. lengths from all-thread up to approximately 12 in.

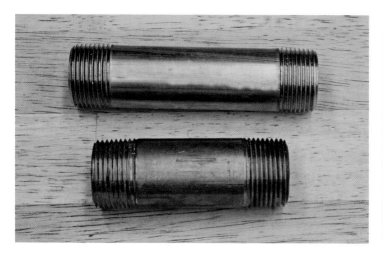

Brass nipples come in both polished (top) and rough (bottom) finishes.

When storing threaded pipe, keep the threaded ends rust free and sharp by wrapping tape around them.

Threaded pipe is an all metal pipe (galvanized or brass) used to connect water lines to fixtures and appliances. Fittings for both materials come in a multitude of angles, as well in short straight sections called nipples. Fittings have threads on the inside (female threads) and nipples and full-length pipe have threads on the outside (male threads). Connecting a fitting to a pipe is accomplished by screwing the threads together. Use pipe wrenches to tighten the connection—one wrench holding the pipe and the other turning the fitting (or vice versa).

>> >> >>

Threaded pipe tips

- Never use a nipple or fitting with rusted threads. Clean the rust off first. Otherwise, the rust will keep the threads from seating properly, causing the joint to leak.
- The bigger the wrench, the easier it will be to tighten the fitting.
- Avoid screwing a metal pipe deep into a plastic bushing or fitting. Plastic cracks easily.
- Always install Teflon tape in a clockwise direction as the threads face you.

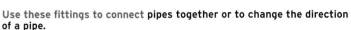

Use these fittings to connect pipes together or to change the direction of a pipe.

Use these fittings to change from one pipe size to another.

Getting a third hand

When tightening and loosening threaded pipe, you'll often need a third hand, and that comes in the form of a pipe vise or a chain vise. These are worth their weight in gold when you need them.

A chain vise is lightweight and will have to be bolted onto something. A pipe vise is heavy and often can be used without being fastened down. Most spin 360 degrees and will tilt in any direction.

A chain vise is lightweight and will have to be bolted onto something.

A pipe vise has teeth shaped into a V that grab the pipe. It is heavy and can often be used without being bolted down.

THREADED PIPE AND FITTINGS (CONTINUED)

A dresser or compression **coupling, made to splice galvanized pipe, should be avoided because it tends to blow off on the pipe.**

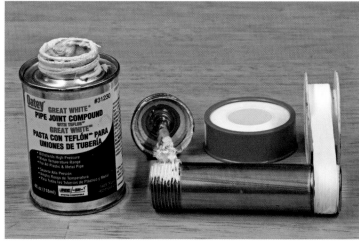

Pick a thread lube: **either paste (left), often called "pipe dope," or Teflon tape (right).**

Splicing galvanized pipe

Always work with threaded ends when splicing galvanized pipe, even though a fitting called a dresser or compression coupling (a metal fitting with rubber gaskets at both ends) is made to do just that. Avoid these fittings if at all possible because they tend to blow off the pipe no matter how much you tighten down the end caps.

Lubricating the threads

To keep pipe threads from rusting and seizing once assembled, coat the male threads with a thread lubricant—pipe dope or Teflon tape. These coatings reduce friction, allow deeper seating of the pipe in the fitting, prevent leaks, and let you unscrew the joint more easily, if needed. Each has advantages and disadvantages.

Although all you have to do with a paste is brush it into the threads, it's messy, so bring a rag and wipe off any excess. Teflon tape must be wound in a clockwise direction (as you face the pipe end), and even when wound correctly, it can tend to pull itself into a useless string. Winding the tape in a counterclockwise direction will cause it to unwind off the threads as you install the fitting.

Teflon tape has a nasty **habit of pulling itself into a string, making it useless. Cut it, throw it away, and start over.**

The easiest way to remove **rust from threads is with a wire brush on a grinder.**

INSTALLING A THREADED FITTING

Don't even think about threading fittings on a pipe unless both male and female threads are clean. Clean heavy rust on male threads by hand, using a wire brush ❶. Alternatively, use a bench-mounted grinder (bottom photo, facing page). If the pipe is in a fixed location, use a wooden-handled wire brush or mount a circular brush on a drill. Apply thread lubricant (paste or Teflon tape) to the male threads before screwing the fitting onto the pipe ❷. Set the fitting against the threads square to the pipe. Turn the fitting clockwise until the threads engage.

Keep turning the fitting until it's hand tight ❸. To get the fitting watertight you must use two pipe wrenches and significant muscle power. Use one pipe wrench to turn the fitting clockwise and a second wrench on the pipe, applying equal power in opposite directions ❹. The jaws should face each other. Tighten the fitting until its orientation stops and is lined up with the pipe on the other end. If you go too far, don't back the fitting off, continue tightening until the faces of the fitting and the pipe meet.

WHAT CAN GO WRONG

Any debris that breaks loose on the inside of a galvanized pipe can clog up all the faucets in the house. Be sure to clean the threads thoroughly, and wipe the inside of the pipe clean with a rag.

1 If the threads are rusted or dirty, clean them with a wire brush and a rag.

2 Spread pipe dope on the threads (or wind on Teflon tape clockwise).

3 Turn the fitting clockwise until it's hand tight.

4 Using opposing pipe wrenches, tighten the fitting onto the pipe.

PUSH-ON FITTINGS

Push-on fittings are a gift from heaven for both the pro and the do-it-yourselfer alike. They are redefining the way we plumb houses. With them, labor to install a fitting has decreased to about 2 seconds for copper, CPVC, and PEX. Average that out for an entire house, and your cost savings become substantial.

Push-on fittings are available in two types: brass and plastic. Brass push-on fittings go by the name of Sharkbite® and are sold in many plumbing supply houses, hardware chains, and on the Internet. Simply push a pipe with a square, clean end into the fitting until you feel it hit an internal stop—about 1 in. into the end of the fitting.

Push-on fittings are perfect for switching from one pipe material to another. The plastic fittings are installed the same way and are available at most large plumbing supply houses.

Brass push-on fittings **come in all the required configurations to completely plumb a house.**

Plastic push-on fittings **work identically to the Sharkbite metal push-on fittings. Some plastic designs have a nasty habit of leaking, however.**

Fitting tips

- Before you install the push-on fitting, always check the pipe end to make sure it has been cut straight and is free of debris and sharp edges.
- Never reuse a damaged fitting.
- Never be in such a hurry that you don't know if the pipe is fully inserted.
- Always feed the pipe straight in. Do not put side or angled pressure on the pipe where it enters the fitting.

Choosing the right push-on fitting

Though all push-on fittings are allowed by many codes, some municipalities have restrictions on their use. Most plumbers prefer all-brass over all-plastic fittings because all-brass push-on fittings (such as Sharkbite) rarely have a problem. Plastic fittings are still carrying the stigma of the failure of PB fittings. And even though manufacturers claim the problem is solved, it may be many years before these fittings are fully accepted. Be sure to check your local codes before you choose between brass and plastic.

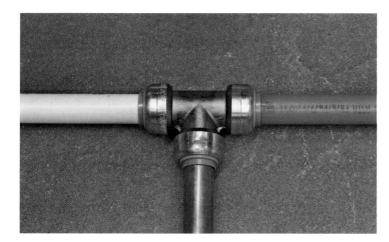

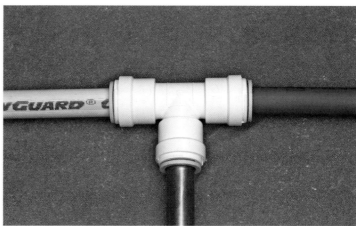

Both metal and plastic push-on fittings, whether T-fittings (as shown here), couplings, reducers, and all others, can accommodate CPVC, PEX, and copper pipe.

These little orange horseshoe pipe-removal tools are available wherever Sharkbite fittings are sold. They also work on plastic fittings.

REMOVING A PUSH-ON FITTING

In case you change your mind about what pipe you want where, push-on fittings won't lock you in.

You can use a special horseshoe-shaped device to remove the fitting. The device is marketed for the Sharkbite fittings but will also work for plastic models. However, the tool can be purchased only where Sharkbite fittings are sold.

Simply snap the pipe-removal tool over the pipe, press it hard against the fitting, and twist out the pipe.

WHAT CAN GO WRONG

When you work with any pipe, and especially when you work with push-on fittings, be sure to remove any sharp or raised edges on the cut end. Pros might use a reamer, but the average homeowner can use plumber's cloth or common sandpaper. Not smoothing the edge might mean the pipe won't fit into the fitting or, worse, that it fits but cuts the O-ring inside.

INSTALLING A SADDLE VALVE ON COPPER PIPE

Saddle valves allow you to tap into a water line without cutting the pipe and inserting a T-fitting. Their only drawback is their small size severely limits the amount of water that can flow through them. Thus their use is limited to fixtures, such as ice makers and water filters, which do not require large volumes of water.

Saddle valves have a sharp tip which can puncture a copper water line (do not use them on plastic water lines). A small-diameter water line is then attached to the valve with a compression fitting. The advantage of using a saddle valve on copper pipe is that it can be installed with the line fully pressurized. As you puncture the line, water will spray out—but it is quickly stopped as you tighten down the valve handle.

Start by placing the rubber gasket on top of the pipe ❶. Arrange the saddle-valve body over the rubber gasket with a bolt on each side ❷. Assemble the nuts and bolts, finger tight ❸. Make sure the position of the valve is correct, then tighten down the nuts with a wrench ❹. Turn the handle clockwise. This will push the sharpened tip through the pipe wall ❺. You'll feel a sudden give in turning the handle when the tip has punctured the pipe. Go about a half turn more, then back the handle off to release the water from the main pipe, letting it flow through the valve. If you haven't installed the outlet pipe already, turn the valve back off and do so now.

The hardened needle tip can puncture a copper pipe, and its shape regulates the water flow.

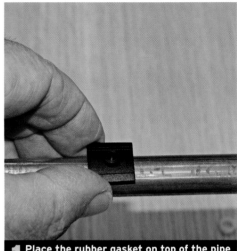

1 Place the rubber gasket on top of the pipe where you want the tap to be.

2 Place the saddle valve over the gasket with the tip centered in the gasket hole.

3 Assemble the valve with the bolts and nuts, and hand tighten them.

4 Square the valve to the pipe, and tighten the nuts securely with a wrench.

5 Turn valve clockwise so the tip punctures the pipe. Then back it off slightly.

INSTALLING A SADDLE VALVE ON GALVANIZED PIPE

When installing saddle valve on galvanized pipe, shut off the water, drain the pipe, then drill a 3/16-in. hole in the pipe. To keep the drill bit from sliding off, tape a small piece of wood to the surface ❶. Once you've drilled the hole, remove the piece of wood and clean off any burrs on the hole with a file ❷. Set the rubber gasket over the hole and assemble the valve, making sure everything lines up properly ❸ ❹. Tighten the bolts, attach the water line, and regulate the flow by turning the handle ❺.

1 Tape a thin board on the pipe to keep the drill from slipping off.

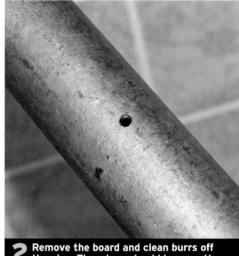

2 Remove the board and clean burrs off the pipe. The edges should be smooth.

3 Place the rubber gasket over the hole.

4 Assemble the saddle valve over the hole and tighten the nuts.

5 Attach the water line and regulate the flow by turning the valve handle.

REPAIRING AND TAPPING WATER LINES

Knowing how to repair or tap into existing water lines—copper, CPVC, PEX, and galvanized—is mandatory if you are going to be doing your own plumbing.

Before you repair or tap into any water line, you have to remove the pressure from that line by turning off the main water valve to the house and draining out as much in-line water as possible. Drain the water out by opening a valve below the line you're cutting. It's also a good idea to turn the water heater off as well. If the water heater is above the line you are cutting, turn the water heater cutoff valve off too, so it can't siphon backward and down into the section of pipe you're working on.

Fixing copper pipe

If you want to stick with copper fittings when repairing or tapping into copper pipe, you will have to completely drain the system so you can sweat on new fittings. It all comes down to a question of whether you want to do it the hard way or the easy way. The hard way relies on the old tried-and-true method of working with copper pipe—cut it, clean it, flux it, solder it. This method is dangerous, a trade skill is required, zero water can be in the pipe, it takes several minutes, and sometimes the solder doesn't sweat properly, especially when you're working in an awkward location. Thanks to the advent of push-on fittings, however, the whole process of installing fittings takes seconds, there can be residual water in the pipe, and no trade skill is required.

➡ See "Push-On Fittings," p. 40.

Fixing CPVC pipe

With CPVC, if you want to make a glue-on repair, you will have to completely drain the system and wait several hours for the glue to dry before you repressurize the system. This is the standard fix. It's the cheapest and takes no special tools.

Using push-on fittings is easier but a bit more expensive. This is the way to go, how-

This is what you'll need **to cut and solder a T, couplings, or other fittings into a copper line the hard way.**

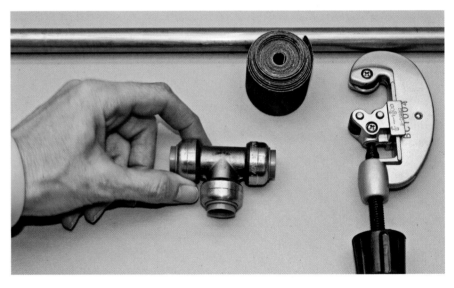

No sweat: This is what you'll need **to repair or cut a T into a copper pipe the easy way, with a push-on Sharkbite fitting.**

ever, because the time savings will outweigh your material costs. Another advantage is that you can remove push-on fittings if you want to drain the line in the future.

➡ See "Removing a Push-On Fitting," p. 41.

Fixing PEX pipe

For PEX, if you want to cut into and cap a PEX line using PEX crimp-on fittings, you do not have to completely drain the line after it has been depressurized. You can crimp a cap onto the line with residual water in the pipe. Using push-on fittings, however, doesn't require special tools. And you can have residual water in the lines. Even a small amount of water dripping out of the pipe, won't interfere with the installation of push-on fitting.

CPVC repair the hard way. You can use CPVC glue and primer, when necessary, to splice or repair CPVC.

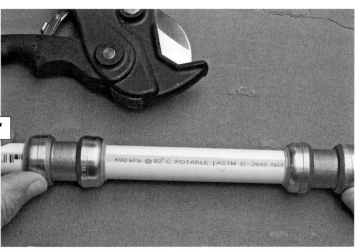

CPVC repair the easy way. A CPVC scissors and two push-on fittings make repairs a snap.

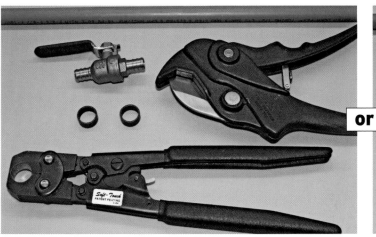

Cutters, crimpers, bands and a valve will let you tap a valve into a PEX line the hard way.

Here's all you need to tap a push-on valve into a PEX line.

Three types of push-on T-fittings for tapping copper pipe the easy way: brass, plastic, and a brass-and-plastic combo.

TRADE SECRET

No matter what kind of pipe you're repairing, always make sure that the cuts you make are exactly square to the pipe. Remove all burrs and rough edges resulting from the cut. These precautions will save you time, and you'll avoid the frustration of having to undo and redo a leaky joint.

REPAIRING COPPER PIPE

Copper sometimes gets dented or kinked, even after it's installed, and it can split when frozen. Fixing the pipe means cutting out the damaged section and replacing it. The key to this process is to cut on either side of the area needing the repair. The procedure shown here uses push-on fittings, but of course, you can install soldered copper couplings if you want to.

→ **See "Sweating Copper Pipe," p. 24.**

The pipe must be perfectly round to accept a new fitting. Make sure you mark the cut at a section where it's not damaged ❶. You can use a hacksaw to cut the copper, but it leaves jagged edges you'll need to file down. A tubing cutter makes cleaner cuts ❷. Make sure you cut perpendicular to the pipe so it's square on the end. Smooth the inside of the pipe with the reamer on the cutter. Using new copper, cut a piece equal to or slightly shorter than the damaged one ❸. Be sure to subtract for any length that will be taken up by the couplings. Insert the replacement pipe into two push-on fittings ❹.

TRADE SECRET
The most common error made when repairing damaged pipe is not cutting far enough away from both sides of the damage. Be sure to cut the pipe where it makes a perfect circle so the fitting will seat properly.

1 Mark cut lines on both sides 1 in. or 2 in. beyond the damaged section.

2 Cut the damaged section with a tubing cutter.

3 Cut a new section of copper to replace the damaged section.

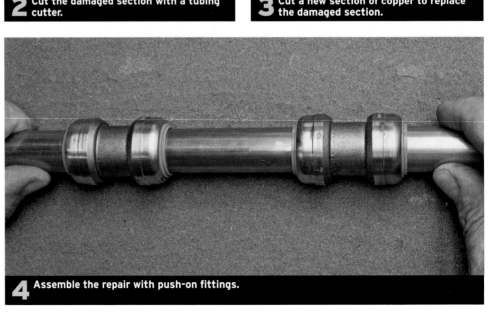

4 Assemble the repair with push-on fittings.

REPAIRING CPVC PIPE

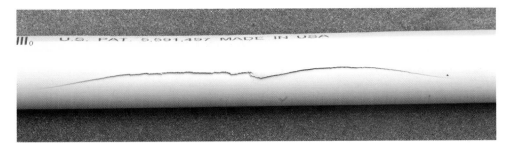

Protect CPVC against freezing, **otherwise it will look like this with water spraying out.**

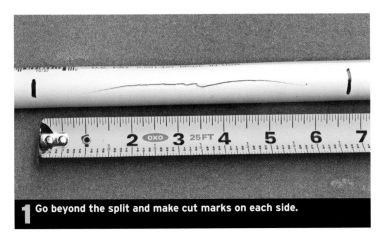

1 Go beyond the split and make cut marks on each side.

2 Cut the pipe at the marks.

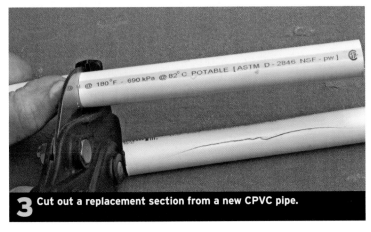

3 Cut out a replacement section from a new CPVC pipe.

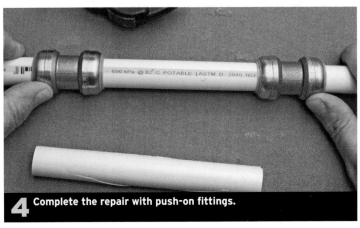

4 Complete the repair with push-on fittings.

Both copper and PEX can take a mild freeze. CPVC cannot. When water freezes inside CPVC pipe, you'll probably be dealing with a burst pipe.

As with other pipes, you must cut out the damaged section. Be sure to make your cuts far enough away on either side of the damage so you're working with perfect non-cracked pipe edges. Push-on fittings make the repair easy (and slightly more expensive than gluing CPVC couplings), but are especially helpful if you're working in a location where it's hard to get the fittings on or you don't want to wait for the glue to dry. Using CPVC glue and couplings takes longer, but is cheaper.

> ➡ **See "Gluing CPVC Pipe," p. 30.**

The problem with CPVC, especially a section that has frozen, is a hairline split you don't see. So look closely. Mark cutlines 1 in. or 2 in. beyond the damaged area **❶**. You can cut CPVC with any fine-tooth blade, but CPVC scissors or a mini-reciprocating saw is best **❷**. If you use a saw, sand any sharp edges smooth. Cut a section of a new pipe for the replacement that is the same length or slightly shorter than the piece you removed **❸**. Be sure to subtract the length required by the couplings. Install the new section with CPVC couplings or push-on fittings **❹**.

REPAIRING PEX PIPE

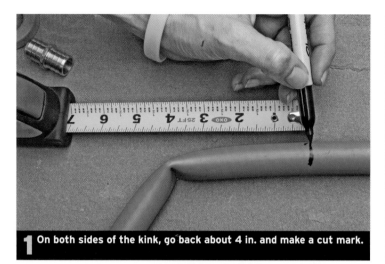

1 On both sides of the kink, go back about 4 in. and make a cut mark.

2 Cut the PEX, using sharp PEX-cutting scissors.

3 Slide a ring on the pipe and insert a coupling. Crimp the ring with a crimping tool.

4 Cut two 4-in. pieces of PEX.

PEX is a sturdy pipe. It's versatile and flexible and will easily make a mild turn. But it won't take a tight turn without kinking. You can tell by working with it how tight to make a turn. If you have to start forcing it, it is too tight. Once kinked (commonly when it goes through a stud or around an object), it will need to be repaired. You'll also have to repair any PEX that's been dented or pierced.

You can use crimped PEX fittings or push-on fittings. Using crimped fittings is not all that hard, but it's easy to make a poor crimp, especially in a cramped position, and it requires an expensive crimper (over $100). Push-on fittings simplify the job.

If the damage lies in a straight run, simply cut it out and install a new piece with two crimped or push-on couplings. Use the same techniques you would use for copper pipe.

> **See "Crimping PEX," p. 34.**

Repairing a kink at a corner or turn will be a little more complicated because you have to crimp on two couplings or add two push-on fittings in addition to the elbow. No matter what fittings you use, first, mark a line on both sides of the damage. Make sure you go beyond the damaged area by several inches so you're working with undamaged and fully round pipe. **1** You can cut PEX with any

sharp knife—even a pocketknife. However, the easiest way is with PEX cutting scissors **2**. Crimp a PEX coupling of the pipe diameter into one of the main pipes. If you can't get the fitting on, the pipe is probably out of round, so cut back even further. **3**. You will need to cut two pieces of pipe to replace the damaged section **4**. Crimp a PEX elbow into one of the replacement sections **5**. Then insert the open end of this section into the coupling in the main line, and crimp both the elbow and the coupling **6**. Repeat the process for the replacement section at other end of the elbow **7**.

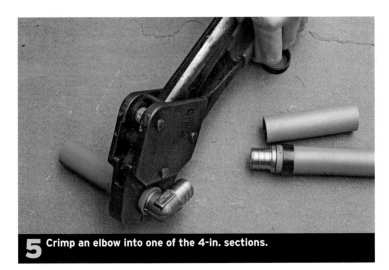

5 Crimp an elbow into one of the 4-in. sections.

6 Assemble the replacement section into the main pipe and crimp the fittings.

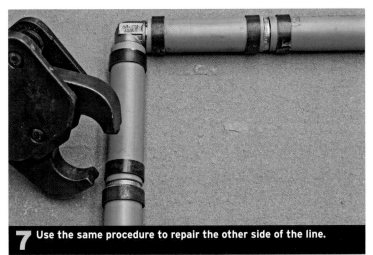

7 Use the same procedure to repair the other side of the line.

Crimping PEX sometimes takes considerable muscle. Here, it took both hands and the entire weight of the installer to crimp the band.

PEX ELBOW OPTIONS

When you're installing an elbow in PEX pipe, you have two options: using a combination of crimp-on and Sharkbite-type push-on fittings or doing the whole job with push-on fittings.

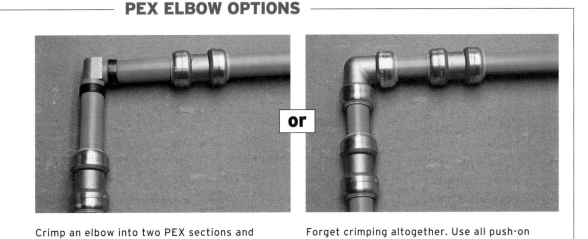

or

Crimp an elbow into two PEX sections and connect to main line with push-on fittings.

Forget crimping altogether. Use all push-on fittings.

REPAIRING GALVANIZED PIPE

1 Cut the pipe several inches ahead of the fitting and threads.

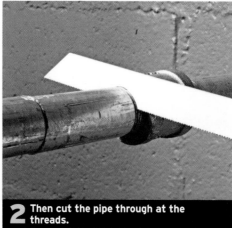

2 Then cut the pipe through at the threads.

3 Use a pipe wrench and a backup wrench to unscrew the coupling from the pipe.

4 Start a female fitting on the threads, then snug the fitting with a pipe wrench.

5 Push the new pipe fully into the female fitting.

If you keep an existing fitting, you'll need a male adapter.

You can repair galvanized pipe or add taps to it just like any other pipe. If you're replacing a union with a T-fitting, you won't have to cut the pipe, but if there's no union, or you're repairing a leak, you will. Cutting galvanized pipe with a hacksaw can take a long time, but a reciprocating saw will reduce the job to a few minutes.

Cut galvanized pipe results in ends without threads, and there's no good way to splice in a replacement or a T. You have to cut out a section of pipe, unscrew the resulting pieces from their respective fittings, and install a new section of non-galvanized pipe. When making changes to a galvanized pipe, you must have threads to work with.

Although leaks sometimes occur in the wall of a galvanized fitting, they are more typical at the threads because the threads reduce the thickness of the pipe wall by more than half.

The type of threads you need for the replacement fitting depends on whether you're discarding the fitting (because it leaks or you're replacing it with a different type) or keeping it. Discarding the fitting leaves males threads on a pipe end that require a female adaptor. Keeping the fitting leaves female threads requiring a male adaptor.

You have to make two cuts: one several inches away from the threads and a second at the threads themselves **1**,**2**. Use two large pipe wrenches to remove the coupling from the pipe **3**. If the coupling is frozen (rusted on), heat it with a propane torch and then remove it. Clean up the threads with a wire brush and rag. Screw on and hand tighten a female fitting appropriate to the replacement pipe. Tighten the fitting with a wrench **4**. Insert the replacement pipe fully into the fitting and tighten it **5**. Follow the same steps at the other end of the new pipe, using a fitting with appropriate threads.

REPAIRING GALVANIZED PIPE

Replacing a Leaky Section of Pipe (or elbow), Discarding the Elbow

1. Cut section of pipe out and discard.

2. Remove leaky elbow and replace.

3. Remove this section of pipe back to first threaded fitting on other end and replace with nongalvanized pipe.

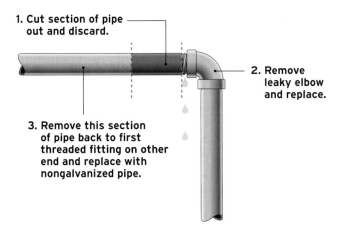

Replacing a Leaky Section of Pipe, Keeping the Elbow

1. Cut section of pipe out and discard.

Elbow is okay, but pipe threads leak.

2. Unscrew stub from elbow. Do not unscrew elbow.

3. Remove this section of pipe back to first threaded fitting on other end and replace with nongalvanized pipe.

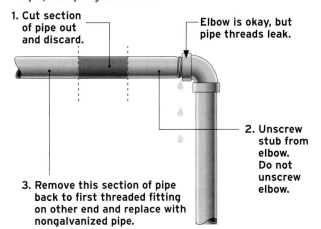

Replacing a Leaky Section of Pipe or Coupling, Discarding the Coupling

1. Cut pipe here and discard cut section.

Male threads

2. Unscrew coupling and discard. Install new coupling.

3. Remove this section of pipe back to first threaded fitting on the other end and replace it with nongalvanized pipe.

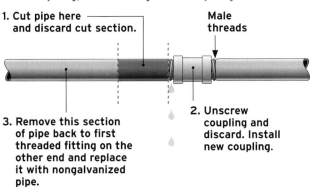

Replacing a Leaky Section of Pipe into a Coupling, Keeping the Coupling

1. Cut section of pipe out and discard.

2. Unscrew stub from coupling. Do not unscrew coupling.

3. Remove this section of pipe back to first threaded fitting on other end and replace with nongalvanized pipe.

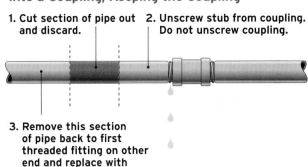

Removing Sections of Leaky Pipe or T-fitting, Discarding the Fitting

2. Cut pipe here and discard cut section.

3. Unscrew T and discard. Replace T-fitting.

1. Cut pipe here and here and discard cut section.

Water leaks from rusted threads.

4. Remove these sections back to first threaded fitting on the other end and replace with non-galvanized pipe.

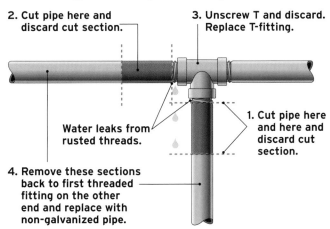

Removing Two Leaky Pipes, Keeping the Fitting

2. Cut section of pipe out and discard cut section.

3. Unscrew stubs.

Water leaks from rusted threads.

4. Remove these sections back to first threaded fitting on other end and replace with nongalvanized pipe.

1. Cut section of pipe out and discard.

INSTALLING T-FITTINGS

You'll need to cut a T into an existing water line when you want to add lines to the supply system. For example, you may want to add a bar sink or perhaps a completely new bath. You can install the T the hard way, of course, soldering copper or gluing CPVC.

➡ **See "Sweating Copper Pipe," on p. 24.**

➡ **See "Gluing CPVC Pipe," on p. 30.**

You can also just cut the pipe and use a push-on T-fitting. You still want to get rid of the water pressure and as much residual water as possible, but any left over water in the lines won't hurt the fittings or slow you down as it would if you were soldering or gluing the pipe. The fittings do not come with marks that show how far the pipe fits into each end. The procedure shown here is the same whether you use copper, CPVC, or PEX.

If wanted, (but not required) you can mark both pipe and fitting for an easier and more accurate installation. Hold the T adjacent to the pipe and mark the pipe where it needs to be cut. This will insure that the cut ends bottom out in the fitting properly ❶. Cut the pipe at the cut marks (be sure to sand down sharp edges) ❷. To make sure the pipe bottoms out fully, hold the fitting parallel to it and mark the pipe at the ends of the fitting ❸. Insert both pipe ends into the fitting so they go in full depth ❹. Insert the tap pipe, in the fitting to its full depth ❺.

1 Mark where the pipe needs to be cut.

2 Cut the pipe at the marks.

3 Mark the pipe at the distance where it should stop in the fitting.

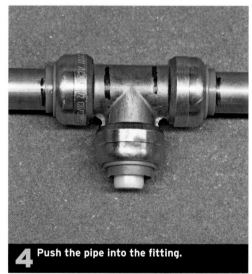

4 Push the pipe into the fitting.

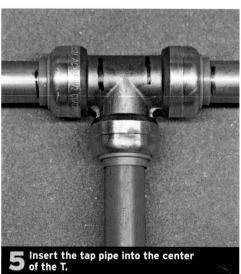

5 Insert the tap pipe into the center of the T.

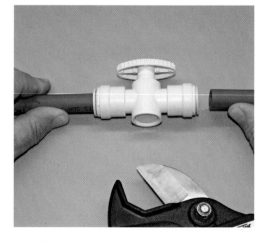

Installing a push-on valve on PEX pipe requires the same easy steps as installing a push-on T-fitting on copper pipe.

CAPPING A PIPE

You'll need to cap a pipe when you want to isolate a section of plumbing from the rest of the house, so the rest of the house can have water while you work on the isolated section.

PEX and CPVC pipe may be flexible enough that you can just cut the pipe, pull it down or to the side and slip on the cap without having to cut out a section **A**. Copper will probably not be that flexible, requiring you to cut a section out to give you some working room **B**. Using a push-on cap makes capping a line fast and easy, and when you want to bring the isolated section back in, just slip the cap back off the line **C**.

A With flexible line, cut the pipe and pull it to the side so you can cap it.

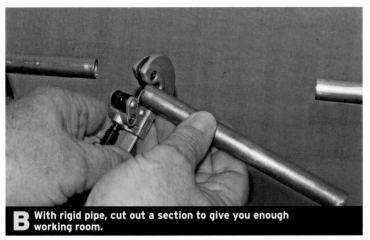

B With rigid pipe, cut out a section to give you enough working room.

C No matter what the pipe material, cap it with a push-on fitting. Remember to sand any sharp edges off the pipe.

Push-on valves

Like other push-on fittings, push-on valves are made to interface with any pipe—copper, PEX, or CPVC. Procedures for installing them are the same as T-fittings. A push-on valve provides an additional advantage capping a line. When you want to stop the water beyond the valve, simply turn it off. Then you don't have to depressurize the system again to tie back into it.

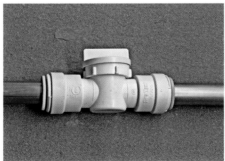

THE DRAIN/ VENT SYSTEM

FOR ALL THEIR APPARENT complexity in design and installation, the pipes in your drain and vent system are not as difficult to work on as they might seem. After all, the system relies on one of the most elemental forces of nature—gravity—to work.

Armed with the proper information, homeowners with average do-it-yourself skills can repair or add to their system without having to call in their favorite plumber.

You will find, however, that your existing drain/vent system is a bit more complex than the water-supply system. That's because there are two things going on at the same time: removing wastewater and allowing air in the lines to accomplish that. Thus, in most cases, there are two sets of pipes for each task—drainpipes and vent pipes—and you'll have to be able to recognize the difference before starting any work on the system.

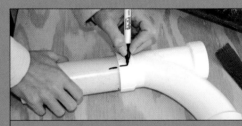

OVERVIEW OF SYSTEM

The drain/vent system carries waste and water out of the house and prevents the entry of dangerous sewer gasses. The primary conduit for all this work is the main vent stack, a pipe usually 3 in. to 4 in. in diameter running through the roof.

One or more secondary stacks of smaller pipe (2 in. to 3 in.) act as branches of the drain/vent system and branch drainpipes, typically 1½ in. to 2 in., carry waste water from fixtures to a stack. Older materials, such as cast iron and galvanized piping, have gradually been replaced by plastic pipe, first a black-colored ABS, then white- or cream-colored PVC. Always check with your local building codes to make sure the materials you're using are approved.

The system of pipes and stacks carries wastewater to the main drain line, where it flows to the municipal sewage system or a septic system. Gravity makes the whole system work, but in order for the water to flow down at the correct velocity, all drainpipes must be sloped at a minimum rate of ¼ in. per foot.

TYPICAL DRAIN/VENT SYSTEM, USING VENT PIPES

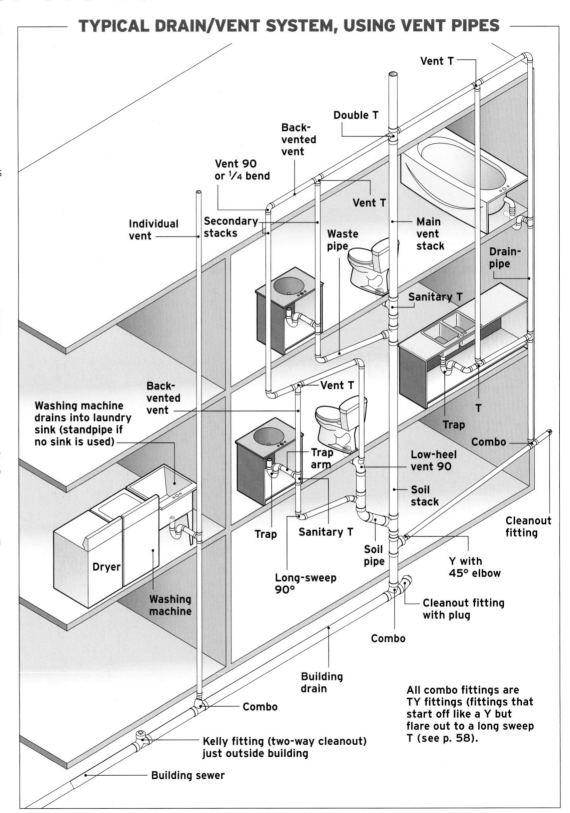

All combo fittings are TY fittings (fittings that start off like a Y but flare out to a long sweep T (see p. 58).

DRAIN/VENT SYSTEM
USING AIR ADMITTANCE VALVES (AAVS)

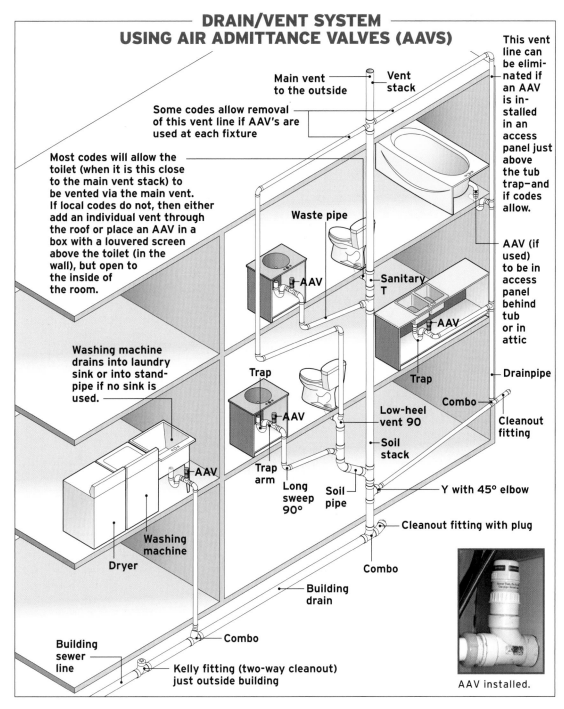

Main vent to the outside

Vent stack

Some codes allow removal of this vent line if AAV's are used at each fixture

This vent line can be eliminated if an AAV is installed in an access panel just above the tub trap–and if codes allow.

Most codes will allow the toilet (when it is this close to the main vent stack) to be vented via the main vent. If local codes do not, then either add an individual vent through the roof or place an AAV in a box with a louvered screen above the toilet (in the wall), but open to the inside of the room.

Waste pipe

AAV

AAV (if used) to be in access panel behind tub or in attic

Sanitary T

AAV

Drainpipe

Washing machine drains into laundry sink or into stand-pipe if no sink is used.

Trap

Trap

Combo

Cleanout fitting

Low-heel vent 90

Soil stack

AAV

AAV

Trap arm

Trap

Long sweep 90°

Soil pipe

Y with 45° elbow

Cleanout fitting with plug

Washing machine

Combo

Dryer

Building drain

Building sewer line

Combo

Kelly fitting (two-way cleanout) just outside building

AAV installed.

Principles of venting

For waste to flow in a drainpipe smoothly it must have an unrestricted air passage in front of and behind it. Otherwise the movement of the water would create a vacuum behind it and high pressure in front of it (pushing air bubbles out the toilet), slowing or actually stopping the flow (and pulling water from the traps). Vent pipes provide this open airway.

Thus all drainpipes in a house must be connected to a vent pipe so that waste can be carried away efficiently without the problem of creating an air pressure wave ahead of the water flow or creating a vacuum behind it. In some cases, drain-pipes are connected directly to a main or secondary stack, which travels through the roof. In others, the drainpipe is joined to a revent which travels up and over to join a stack vent.

Air admittance valves (AAVs) use the natural vacuum created by flowing water to open a valve and let air into the drain line, allowing waste water to flow smoothly. The valve closes immediately after drainage, preventing the introduction of sewer gasses into the room. AAVs can drastically reduce the number of vent pipes needed.

DRAINPIPE AND FITTINGS

Drainpipe and fittings come in two types of plastic: polyvinyl chloride (PVC), which is white, and acrylonitrile butadiene styrene (ABS), which is black. Both are durable and code compliant, and if properly installed, both will outlive the average homeowner. ABS is more popular in the West and PVC is more widely used in the Midwest and East. Where you live may therefore affect the availability and the cost of the materials you use. Each material requires a glue made specifically for it.

Drain/vent pipe comes in both **PVC (white) and ABS (black)** in various sizes to accommodate specific needs of the various parts within the drain/vent system.

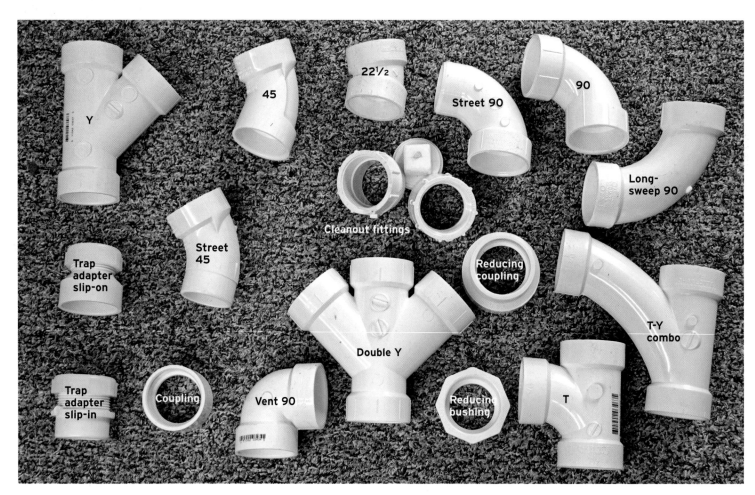

Drain and vent pipe comes in configurations to match the needs of any plumbing project. Shown here are some of the most common (they can be white PVC, or black ABS.) When installing new plumbing, buy an ample supply of fixtures from a dealer who will allow return of unused materials.

Traps

A trap is a fitting designed to hold water at the base of a plumbing fixture, such as a sink, toilet, or shower, to prevent sewer gases from escaping into the room.

Standard traps are less expensive and are the most commonly used, but once glued, they are fixed in place. A union trap allows removal of the trap arm back to the wall but gives you a possible leaky joint.

A trap with a bottom drain allows access to get out lost items, such as a wedding ring or faucet part lost during repair.

Neither the union trap nor the drain trap are commonly used since the standard trap with slip nuts can be easily removed by loosening the nuts at the wall and under the sink.

➔ See "Running Vent Lines," p. 64.

Coupling drain-pipe sections.

To couple drain-pipe sections together, you can use glued or flexible couplings. Glued couplings are inexpensive, but the pipe on both sides of the coupling must lie very closely on the same plane. A flexible coupling is more expensive but allows considerable angle change. Many installers use flexible couplings for a small change of direction or when bringing two ends together so they don't have to worry about getting both pipes on the same plane as they meet.

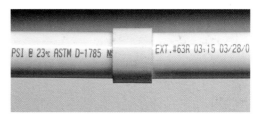

A plastic coupling is the most common method of joining two lengths of pipe.

A flexible coupling allows a small change of direction when two lengths of pipe meet.

HOW A P-TRAP WORKS

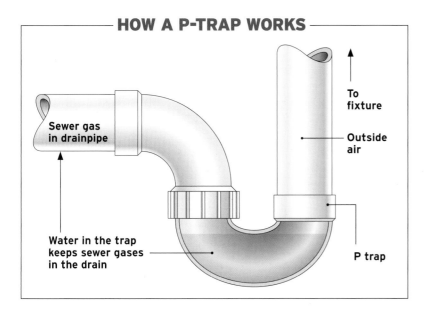

Sewer gas in drainpipe

To fixture

Outside air

Water in the trap keeps sewer gases in the drain

P trap

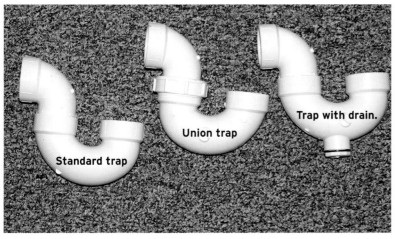

Standard trap

Union trap

Trap with drain.

Traps stop sewer gases from entering the house and are sold with slip fittings and solvent fittings, as shown here.

CHOOSING FITTINGS

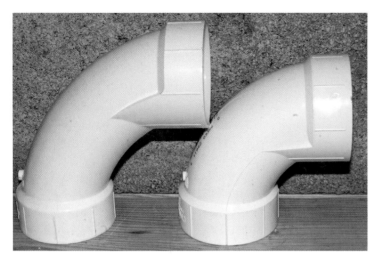

A long-sweep elbow (left) is about 1 in. taller than its shorter cousin (right) and allows for better drainage.

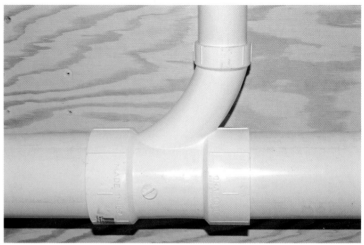

Typically used when a drain is directly overhead, this TY combo fitting angles the fluid into the main pipe run as a Y-fitting.

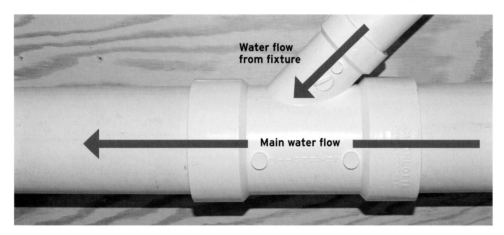

Water flow from fixture

Main water flow

Typically used when adding fixtures to the main drain line, this Y allows fluid to enter at an angle without slowing its movement.

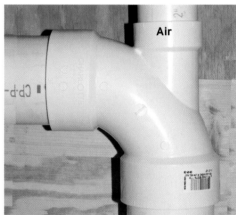

Air

Used primarily to add a vent, a vertical heel-off fitting lets vent air or drain water enter from above.

Plastic drainpipe will come in different thicknesses (grades or "schedules.") For residential plumbing, stick to schedule 40. Various plumbing renovations will require different configurations of drainpipes and vent pipes and fittings, but the photos shown here will give you an idea of the kinds of fittings you should use in different applications. In general, drain fittings (sometimes called sanitary fittings) make gradual turns to reduce the chance of clogs; vent fittings can make immediate turns. A few basic guidelines can help you organize your plan.

Vents

Because vent pipe fittings deal with gasses, they can have sharper bends, and you won't need sweep fittings. All drain fittings that carry fluids do need them. However, you can still use drain fittings if you want. Drain fittings are commonly used for vents for two reasons, especially when you have a large amount of new work. You don't have to stock two different types of fittings, and you avoid the possibility of mixing vent fittings with sanitary fittings and failing an inspection.

Horizontal to vertical drain runs

In the wall behind individual fixtures, use a common drain elbow (90) or T-fitting (1½ in. or 2 in. only). For the main sewer or drain line runs, use a long sweep elbow, a TY-fitting with cleanout, or two 45s separated by a short piece of drainpipe.

Vertical to horizontal drain runs

A 90-degree change in direction means installing fittings with a pronounced gradual sweep.

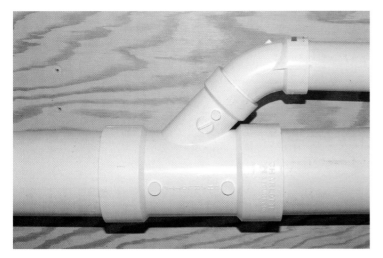

Typically used when the drain is at some distance **from a fixture, a Y-fitting and a street 45 turned parallel to the pipe work well for a horizontal input.**

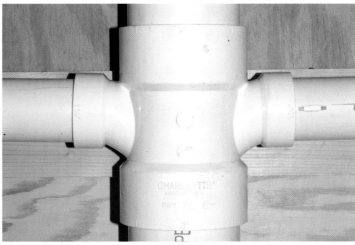

A vertical double-inlet T **allows drain branches to enter from both sides but isn't allowed for back-to-back toilets by some codes.**

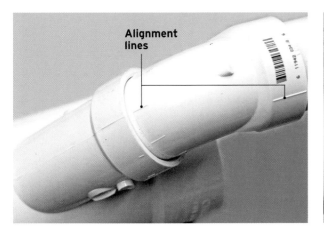

Integral alignment lines **on both fitting and hub help keep the pipes straight.**

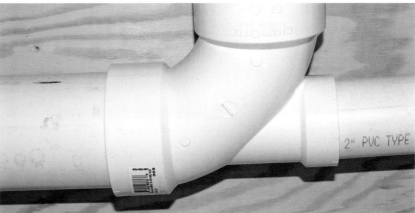

A horizontal heel-off fitting **should not be used for venting but for drain water. Some codes require that the heel pipe only go a few inches before it must turn up.**

The most misused drain fitting is the 90-degree elbow. That's because it can make a tight turn fast. However, this fast turn can create clogs and slow the fluid down to the point that solids start to drop out. Thus, whenever possible, always try to use 45-degree drain fittings separated by a few feet of pipe.

For tying branch circuits into a main line, use Y-fittings as much as possible. After that, use a TY-fitting. You could also use a Y-fitting with a street 45. Use Ts only when you have no other choice. Though they work well, they slow down the water flow compared to a Y-fitting and are against code in some areas.

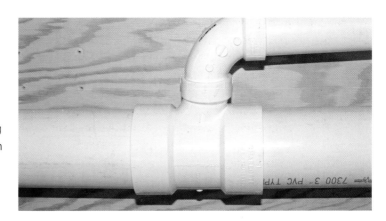

Typically used when there is no room **for a Y or TY, a T-fitting with a street elbow used like a periscope works very well but is against code in some areas.**

CHOOSING THE RIGHT SIZE DRAINPIPE

You must pick a drainpipe diameter large enough to keep the fluid from slowing down (you need at least 2 ft. per second) and to allow air in the pipe at all times. Practically speaking, there are four different pipe diameters to work with: 1½ in., 2 in., 3 in., and 4 in. (stay away from 1¼ in. because its small diameter can become easily clogged and fill with water).

There are very complicated tables you can use to help you choose minimum pipe diameters, but the following simple rules of thumb will keep you out of trouble:

- Use 1½-inch pipe only for short runs from a single fixture and immediately next to a fixture trap, such as a lav or kitchen drain.
- Use 2-in. pipe for long horizontal runs to a single fixture (like the kitchen sink) or two fixtures on a single drain line (called a branch).
- If the kitchen sink is on an isolated island, use 3-in. pipe. For more than two fixtures on a branch, use 3-in. pipe.
- Use 4-in. pipe for the main drain line in the house, typically from the toilet to the septic tank or city sewer line.
- You can run 3-in. pipe to a second toilet off the 4-in. line from the first toilet. If you have three toilets, take 4-in. pipe at least all the way to the second toilet.

For older houses that use 3-in. pipe as the largest pipe in the house, then that is what you will have to live with. Everything may work fine, but you have a much higher chance of clogs and venting problems, especially if you have more than one toilet on the line.

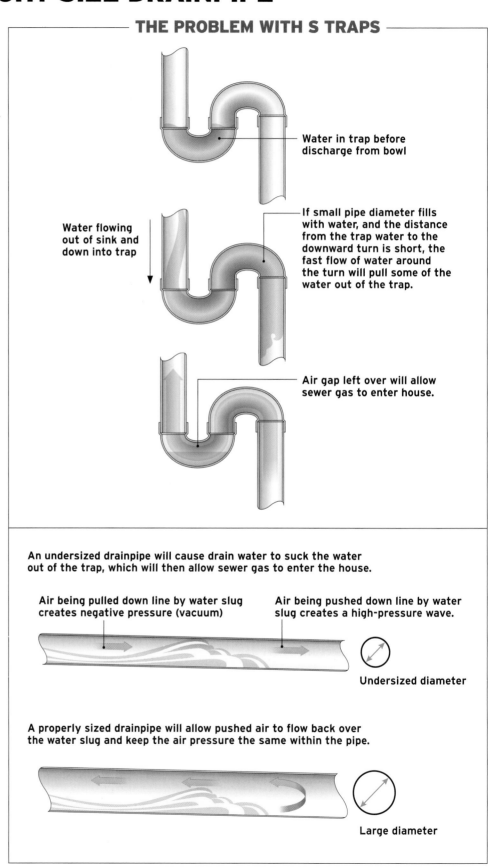

THE PROBLEM WITH S TRAPS

Water in trap before discharge from bowl

Water flowing out of sink and down into trap

If small pipe diameter fills with water, and the distance from the trap water to the downward turn is short, the fast flow of water around the turn will pull some of the water out of the trap.

Air gap left over will allow sewer gas to enter house.

An undersized drainpipe will cause drain water to suck the water out of the trap, which will then allow sewer gas to enter the house.

Air being pulled down line by water slug creates negative pressure (vacuum)

Air being pushed down line by water slug creates a high-pressure wave.

Undersized diameter

A properly sized drainpipe will allow pushed air to flow back over the water slug and keep the air pressure the same within the pipe.

Large diameter

CLEANOUTS

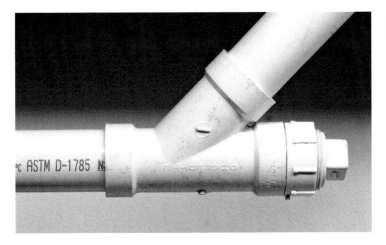

One type of cleanout fitting can insert directly into a fitting's hub when a change of direction occurs.

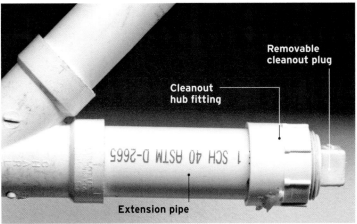

To extend the cleanout to an area with enough clearance to work in, you may have to use a hub type cleanout fitting and an extension pipe.

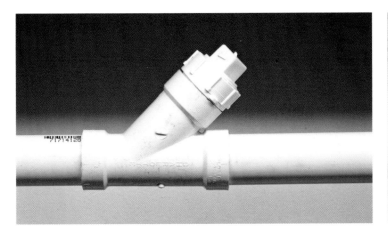

A Y with cleanout plug can be inserted into a straight run but it can only be rodded in one direction.

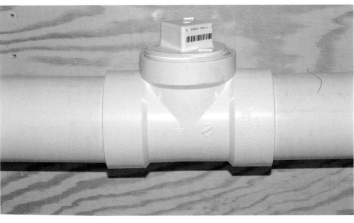

This cleanout fitting design for main drain/sewer lines allows rodding in both directions.

B ecause cleanouts are typically installed to allow rodding (mechanical cleaning) of clogs in the lines, they are installed for direction changes over 45 degrees, at the base of a stack, on horizontal runs over 100 feet, at the end of each branch line, and anywhere you'd logically want to rod a clogged line. Always install at least two cleanouts in the main drain/sewer line (the line the toilet is on). Place one where the line enters the house (preferably outside). Install the second at the end of the run to the toilet. This line can extend to the outside, or terminate under the floor, or extend into an area upstairs. However, the cleanout must be accessible—you'll need at least 18 in. between the plug and a wall or other physical obstruction to have room to work with rodding equipment.

➡ See "Rodding a Drain Line," p. 92.

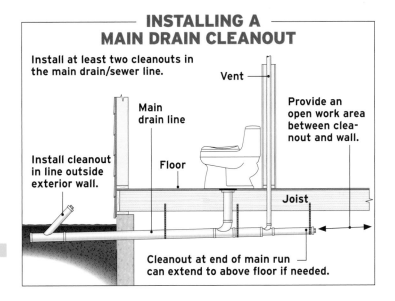

INSTALLING A MAIN DRAIN CLEANOUT

Install at least two cleanouts in the main drain/sewer line.

Vent

Main drain line

Install cleanout in line outside exterior wall.

Floor

Provide an open work area between cleanout and wall.

Joist

Cleanout at end of main run can extend to above floor if needed.

RUNNING VENT LINES

The purpose of a vent line is to prevent the water in the traps from being pulled out by a vacuum caused when rapidly running waste water flows down a pipe.

Vent line theory and technology have evolved substantially over the years. Early system design called for one vent for every fixture. Then we found out we could have a common vent system for several fixtures and vent them with a single pipe. The latest innovation is the air admittance valve (AAV), which replaces individual vent lines running through the roof.

AAVs are accepted by all the accredited agencies but are still prohibited in some localities. If codes prohibit their use in your area, you'll have to run an individual vent from a fixture or install drainpipe from the fixtures to a line sized to prohibit the formation of a vacuum. You can also create a very expensive loop of pipe as some do for an island drain in a kitchen. The illustrations on these pages demonstrate a number of solutions to common problems experienced when running vent lines.

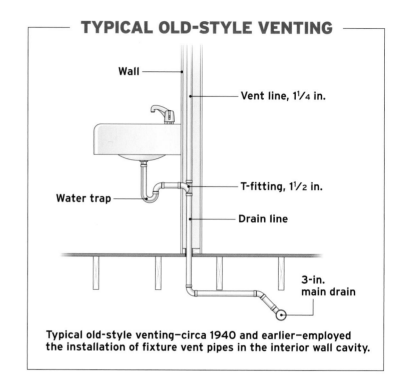

TYPICAL OLD-STYLE VENTING

Wall

Vent line, 1¼ in.

T-fitting, 1½ in.

Water trap

Drain line

3-in. main drain

Typical old-style venting—circa 1940 and earlier—employed the installation of fixture vent pipes in the interior wall cavity.

WORKING AROUND OBSTRUCTIONS

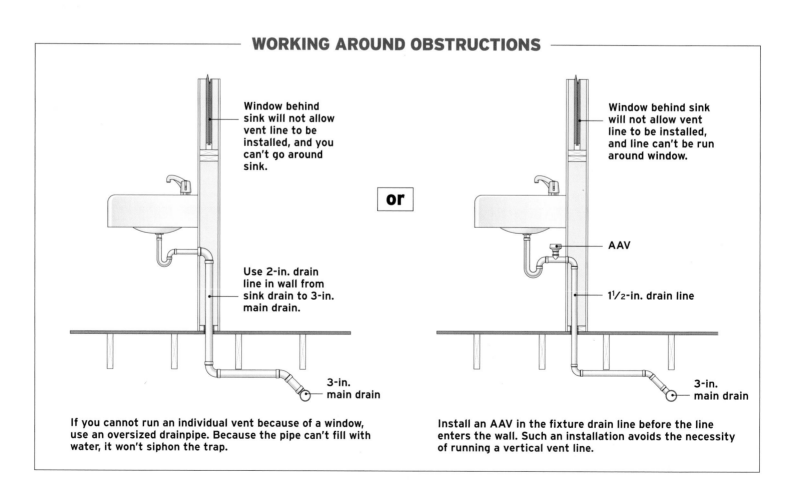

Window behind sink will not allow vent line to be installed, and you can't go around sink.

Use 2-in. drain line in wall from sink drain to 3-in. main drain.

3-in. main drain

or

Window behind sink will not allow vent line to be installed, and line can't be run around window.

AAV

1½-in. drain line

3-in. main drain

If you cannot run an individual vent because of a window, use an oversized drainpipe. Because the pipe can't fill with water, it won't siphon the trap.

Install an AAV in the fixture drain line before the line enters the wall. Such an installation avoids the necessity of running a vertical vent line.

INSTALLING A FIXTURE VENT

Before running a vent line horizontally, the vertical vent section must travel at least 6 in. above the flood rim of the fixture it is venting. This keeps sludge from settling in the vent line in case of a backup.

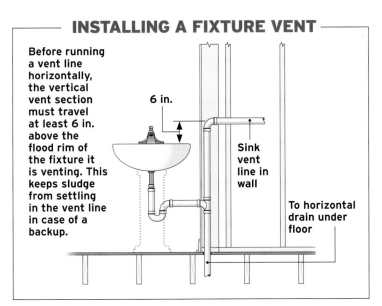

6 in.

Sink vent line in wall

To horizontal drain under floor

VENTING WITH AN INVERTED T-FITTING

Install a sanitary T-fitting upside down to allow the vent gas to flow upward and out the vertical vent more easily.

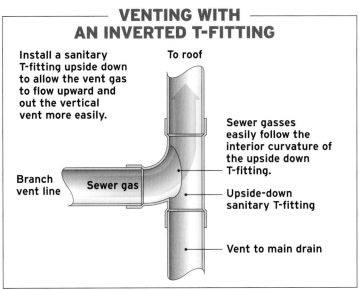

To roof

Sewer gasses easily follow the interior curvature of the upside down T-fitting.

Branch vent line

Sewer gas

Upside-down sanitary T-fitting

Vent to main drain

VENTING IN JOIST SPANS

Venting within the Joist Span

Use a street 90 to point the pipe to the nearest wall to make it easy to vent a drain line.

Floor

Vertical vent to attic

Street 90 glued into top of T-fitting

90° elbow

T-fitting

Centerline of sewer pipe

Venting outside the Joist Seam

Extend the inlet pipe and use two 90s to get to the vent wall.

90° elbow

AVOIDING VENT CLOGS

Never take off for a vent air beyond 45° from vertical. Below 45° from vertical, a vent line can plug up with water and debris.

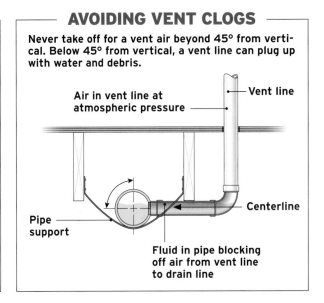

Air in vent line at atmospheric pressure

Vent line

Centerline

Pipe support

Fluid in pipe blocking off air from vent line to drain line

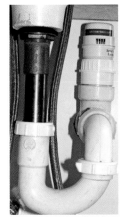

Instead of running expensive **revent loops**, install an AAV in the drain line at the fixture.

Large AAVs can vent an entire branch (a pipe off the main line going to several fixtures) at one time.

GLUING A DRAIN FITTING

When measuring the length of a pipe you need between fittings that are already installed, always measure the distance from the inside of one hub to the other. Dry-fit the pieces before you measure so you don't come up too short or long.

Once you know how long a pipe section you need, cut the pipe to length using the procedures shown here. You can also use a PVC saw, reciprocating saw, handsaw, miter box, or hacksaw instead of a cut-off saw.

Mark the pipe at the correct length and position it on the table of your cut-off saw, square to the blade and lined up so the blade cuts to the waste side of your marking. Turn the saw on and pull the blade down to make the cut ❶. Cutting plastic pipe normally leaves raised nubs of rough plastic. Using plumber's cloth, sand the cut area to remove any edges and leftover pipe ❷. Place the pipe against the hub of the fitting and mark the hub depth on the pipe ❸. This mark will tell you how deep the pipe needs to be inserted into the hub. Dry-fit pipe and fitting together up to the insertion mark ❹. Add alignment marks if needed so the fitting will point the right direction once it is glued ❺. Apply primer to both pipe and fitting ❻. Be sure to use a large diameter dauber (about the size of a quarter) for the larger drainpipe. Apply glue to both pipe and fitting ❼. Insert pipe and fitting together ❽. Make sure the pipe is inserted all the way into the hub. Turn the pipe or fitting to proper angle if needed to align it to a specific direction or mark ❾. Hold for around a minute to keep the two together—otherwise they will pull apart.

1 No matter what kind of cutting tool you use, always keep the pipe square to the blade and cut pipe to the waste side of your dimension line.

4 Dry-fit the hub and pipe, making sure the pipe will fit in the hub up to the edge of your depth line.

7 Using a large dauber, apply glue to the pipe and fitting. Use ABS glue for black pipe and PVC glue for white pipe.

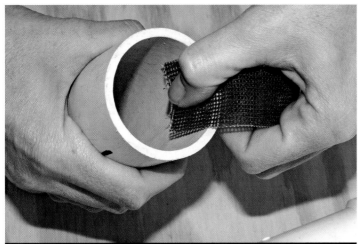

2 Using open or closed plumber's cloth or a sanding screen, smooth the cut edge of the pipe.

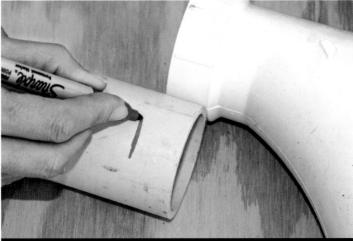

3 Mark the depth the pipe will need to be inserted into hub. Remember which side of the line represents the waste side.

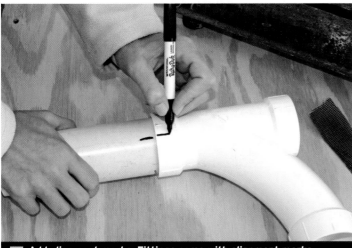

5 Add alignment marks. Fittings come with alignment marks, but you may want to mark over them to make them easier to see.

6 Apply primer to both the outside of the pipe and the inside of the fitting. Black pipe (ABS) does not need a primer.

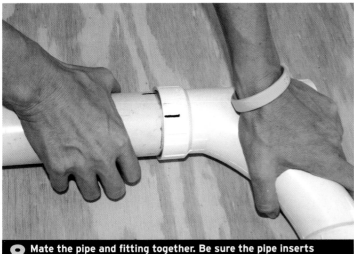

8 Mate the pipe and fitting together. Be sure the pipe inserts all the way into hub.

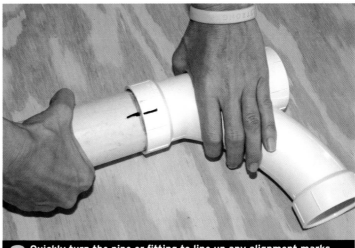

9 Quickly turn the pipe or fitting to line up any alignment marks. Hold the pieces in place for a minute to let the glue set.

HANGING DRAINPIPE

HANGING PIPE PROPERLY

A A single wrap of metal strapping is the most common way of hanging pipe.

B For more secure support, wrap the hanger strap around the pipe.

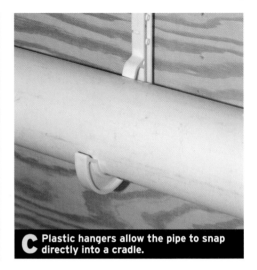

C Plastic hangers allow the pipe to snap directly into a cradle.

Horizontal drainpipe has to be supported on hangers hung from the overhead joists approximately every other joist. Going too far between hangers will cause the pipe to sag between hangers over time, creating an uneven path for the wastewater. Low sections of pipe will accumulate debris that falls out of the water, eventually clogging the drain line.

Hanging drainpipe is done in two stages: First hang the pipe temporarily on its

intended path and dry-fit any fittings to make sure everything is oriented properly. Typically you will have to adjust the pipe and hangers up and down and left and right to get the proper orientation and slope (1/4 in. per running foot). Once you have positioned the pipe exactly where you need it to be, screw or nail the hangers in place.

You can hang pipe from plastic or metal snap-on hangers or simply hang it from rolled low-cost perforated metal

strapping **A**,**B**,**C**. Do not use rolled plastic hanger strap—it will stretch over time and possibly snap.

Don't guess when hanging the pipe at the required 1/4-in.-per-ft. slope. Use a 4-ft. level and place it against the drainpipe. Lower the upper end 1 in. (cut a 1-inch wood spacer to make this easy). A properly sloped pipe will make the level read plumb (bubble centered in the vial). You can also use a 2-ft. level and lower the upper end 1/2 in.

GETTING THE RIGHT SLOPE

To obtain a 1/4-in.-per-ft. slope, use a 4-ft. level and allow for a 1-in. gap on the upper side between the level and the pipe when level is plumb. Arrange the hangers to create a downhill slope on the drain the pipe of 1/4 in. per foot or 1 in. every four feet.

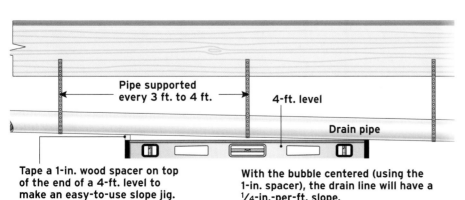

Pipe supported every 3 ft. to 4 ft.

4-ft. level

Drain pipe

Tape a 1-in. wood spacer on top of the end of a 4-ft. level to make an easy-to-use slope jig.

With the bubble centered (using the 1-in. spacer), the drain line will have a 1/4-in.-per-ft. slope.

WHAT CAN GO WRONG

You can't get there from here. It's possible that the toilet flange locator pipe is behind an obstruction and you cannot get a drainpipe to it in a straight line. In this case, you have to use a combination of fittings to get around the obstruction. Make a workaround from small angle fittings such as 45-degree and 22 1/2-degree fittings with several feet of pipe between; never use a pure 90 or elbow unless you have no other choice.

USING FLEXIBLE FITTINGS

Flexible fittings have all the physical characteristics of a rubber material—they look like rubber, they smell like rubber and they're pliable—but they're not rubber. They're made from flexible PVC, and that flexibility can help bail you out of otherwise insolvable problems and make some routine chores easier.

A few of their applications are shown here, but like those of us in the trade, you'll probably think of countless more. For example, they work well in hooking up an entire cast-iron system, for those who want really quiet plumbing. They'll help you fix places where a straight pipe isn't quite at the right angle for a fitting. The flex fitting will bring them in line. For cleanouts, you can take an elbow right off the line and easily rod downstream or upstream.

Installing a flexible fitting **took less time than measuring and gluing a solid PVC fitting, and allows some "wiggle" room.**

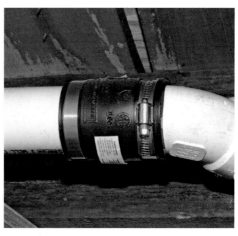

Flex fittings **solve alignment problems in a flash.**

A flex elbow-trap **will "give" with movement of the tub or shower and won't pressure the drain pipe or fitting into cracking.**

Flex fittings **can absorb vibrations in the pipes, making the system run more quietly.**

BRINGING PIPES IN LINE

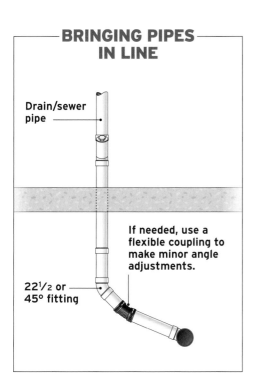

Drain/sewer pipe

If needed, use a flexible coupling to make minor angle adjustments.

22½ or 45° fitting

WHAT CAN GO WRONG

Sometimes at the end of a run, it is extremely hard to get the drain line at the exact angle to glue it properly. In these situations, cut the pipe as you normally would and use a flexible coupling a few feet before the end of the run to change the angle by a few degrees.

CUTTING HOLES FOR DRAIN LINES

A super heavy-duty right-angle drill is needed for large-diameter holes. (Photo courtesy of Milwaukee® Tool)

A pistol-grip right-angle drill weighs less than a super heavy-duty drill but won't cut as quickly. (Photo courtesy of Milwaukee Tool)

A hole saw bit is best for drilling large-diameter holes.

CUTTING DRAIN-PIPE HOLES

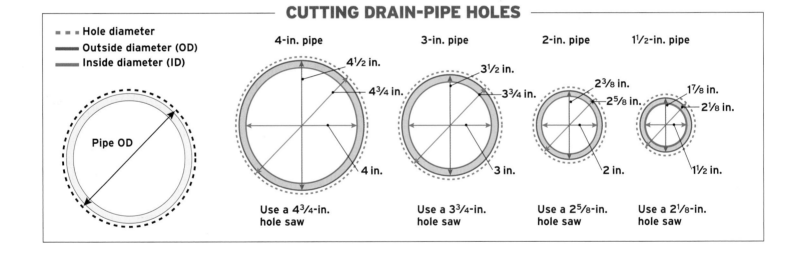

- - - Hole diameter
—— Outside diameter (OD)
—— Inside diameter (ID)

Pipe OD

4-in. pipe 4½ in. 4¾ in. 4 in. Use a 4¾-in. hole saw

3-in. pipe 3½ in. 3¾ in. 3 in. Use a 3¾-in. hole saw

2-in. pipe 2⅜ in. 2⅝ in. 2 in. Use a 2⅝-in. hole saw

1½-in. pipe 1⅞ in. 2⅛ in. 1½ in. Use a 2⅛-in. hole saw

Drainpipes require very large diameter holes which can reduce the structural integrity of whatever they pass through (joists, for example). Thus wherever possible, it is always best to hang the main drain/sewer pipes below the joists instead of drilling them.

When you have to drill holes for the pipes, and the structural integrity is not going to be compromised (in non-weight-bearing walls, for example), you need a right-angle drill equipped with a hole saw bit.

Common pistol-grip drills do not have the torque needed for such large holes. The bit will often bind in the wood, causing the drill body to torque violently in the opposite direction. Drilling with them is quite dangerous, and the torque could break your wrist.

To determine what size hole saw bit you need, use this rule of thumb: You'll want to drill a hole ¼ in. larger than the outside dimension (OD) of the pipe.

To determine the pipe OD, you can measure across the center of the pipe, but this is usually inaccurate. Just add ½ in. to the pipe size, and you'll be on the money or very close. For example, a 3-in. pipe will have an OD of around 3½ in. and will require a 3¾-in. hole saw bit.

You always want a little extra open space around the pipe in the wall, otherwise the pipe will squeak as it expands and contracts.

If you cut a hole within 1¼ in. of the edge of a stud, protect the pipe with metal strips along the stud edge and a plate (where the pipe goes through the floor).

WORKING WITH CAST-IRON AND GALVANIZED PIPE

Once the mainstay of drain lines, interior cast iron is installed today only if you want an especially quiet drainpipe, for example in the walls around dining or living rooms.

Most of the time you'll be working with cast iron and galvanized drainpipe only when you are cutting it out during a remodel.

It used to take special tools to cut cast iron, but today we use diamond blades on a cut-off saw or a circular saw **A**. Diamond blades are also available for reciprocating saws. Galvanized drainpipe can be cut easily with a fine-tooth blade (or diamond blade) on a reciprocating saw **B**.

To interface plastic pipe with cast iron, use a banded Fernco® or a "no-hub" fitting (a flexible coupling with a stainless band around it) **C**. Common flexible couplings without any band protection will also do the job when all allowed by code.

When running cast iron horizontally, you'll need to support the pipe every 3 ft. to 4 ft. Because it's so heavy, you'll also need to support it right after any interface with plastic pipe.

WARNING
Always wear safety glasses, especially when sawing or working overhead.

A Using a diamond blade on a circular saw, cut the cast-iron pipe as much as possible. Finish with a reciprocating saw with a diamond blade.

B For a fast cut in galvanized pipe, use a fine-tooth blade (or diamond blade) and cut the pipe at the threads.

Cast-iron pipe

Support

PVC drain pipe

No-hub fitting or Fernco-style fitting with stainless band.

C Support cast iron immediately after joining it to plastic pipe.

CLEANING AND UNCLOGGING

IF WATER FROM A FAUCET RUNS slower than you think it should, or water drains sluggishly from a fixture, it's probably time for some cleaning and unclogging.

Any number of things—grease, hair, soap scum, naturally occurring sediment, and even toys—can impede the proper flow of water both to and from a fixture. And generally, only a few tools, simple procedures, and a minimum amount of time are all you need to get things running smoothly again.

If a slow-drain problem is limited to only one fixture, the remedy is probably close to that fixture. If several fixtures aren't behaving correctly, look for solutions up or down the drain lines.

Always begin a cleaning or unclogging task with the simplest method first. Start unclogging a drain line, for example, with a plunger. If that doesn't work, use an auger or dismantle the trap. When all else fails, it's time for the heavy equipment—a sewer tape or rodding machine.

SIMPLE CLEANING TASKS

CLEARING BLOCKED LINES

CLEANING A FAUCET AERATOR

Faucet aerators mix air with the water, minimizing splashing and reducing the amount of water used (and the energy required to heat hot water) without reducing the effectiveness of the water stream. An aerator contains a screen and a water reducer/aerator washer, and these little items have a habit of collecting bits of naturally occurring mineral sediment in the water. What you'll notice is a reduced water flow at the spout (on both hot and cold) and/or a nonsymmetrical spray coming from the spout.

To remove the aerator from the faucet simply turn it counterclockwise. Drop it straight down so you don't lose any internal parts, especially the thread-sealing gasket. The threads can be either inside or outside the cap. If the cap is stuck, you will need piers to turn it (tape the jaws with electrical tape to minimize scratching). ❶ Look inside the center area for sand, silt, and other waterborne debris ❷. Take the center section out to check for further debris, noting the order in which things come apart ❸. Check for anything stuck in the screen ❹. In the flow reducer, look in the tiny side holes ❺ and the center hole of the white button for debris ❻. If you do not put all the pieces back together properly, there will be a leak or the water flow will not be a smooth aerated flow ❼,❽,❾.

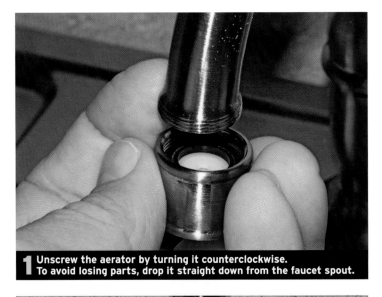

1 Unscrew the aerator by turning it counterclockwise. To avoid losing parts, drop it straight down from the faucet spout.

4 Check the screen for debris and flush it or remove debris with a toothpick or other pointed tool.

CLEANING SCREENS

Sometimes you can remove whatever is clogging an aerator by simply flushing the screen or reducer with water, turning the parts under the spray to dislodge the sediment. Soaking the parts in vinegar can also help dissolve calcium deposits. For really stubborn bits of debris, try cleaning the holes with a toothpick, large sewing needle or other small, sharp tool.

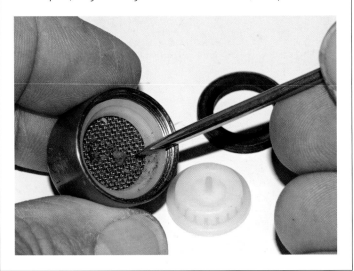

8 Then the aerator disk.

2 Check the central white area for sediment and flush it out.

3 Pop out the center section, and you will get a screen and gasket. Keep track of the order in which the parts come out of the aerator body.

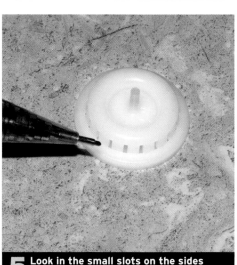

5 Look in the small slots on the sides of the aerator disk and clean them.

6 Check both front and rear of the aerator. and wipe debris clear of the disk.

7 Reassemble by putting the parts back in order: screen first.

9 And finally, the seal.

Cleaning tools. **Though a pick-and-hook set is best for cleaning small parts, you can improvise with tiny screwdrivers or needles. A pick-and-hook set however, will have the little turned ends that work perfectly to get washers and screens out.**

WHAT CAN GO WRONG

Despite your best intentions, it's easy to let the parts of an aerator fall out when you're removing it. To prevent permanent loss of any parts, put the stopper in the sink drain before removing the aerator. If you take the aerator away from the sink, to keep from losing parts, you can disassemble it over a bowl.

CLEANING A KITCHEN SPRAYER

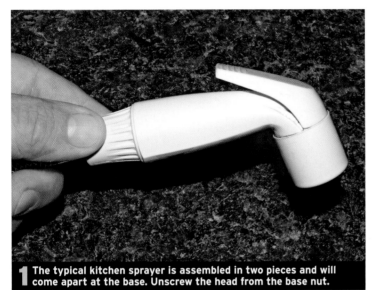

1 The typical kitchen sprayer is assembled in two pieces and will come apart at the base. Unscrew the head from the base nut.

2 Once separated you will see a seal and a spacer on the hose end of the sprayer.

5 Remove the retaining ring with a miniature screwdriver or the point of a knife.

6 Slide the base nut off the hose and soak all the parts in a cleaning solution.

Commonly used to spray the dishes in the sink, the sprayer is quite a busy little tool. Sometimes the control lever breaks or sticks, and sometimes you may want to change out the sprayer entirely to match a different kitchen color scheme. Cleaning or changing the sprayer requires removing it.

Pull the sprayer as far as possible out of its receptacle hole. If it comes out only a short distance, it's probably hung up on the pipes under the sink. In this case, untangle the hose under the sink so it's clear of obstructions.

Note the top and bottom separation line of the sprayer; this is where it comes apart. To disassemble, start by unscrewing the top from the base. Hold the bottom and twist the top toward you **1**. Lay the top aside and pull loose the gasket and spacer **2**. Note the gasket (black) is on top **3**. A U-shaped retaining ring holds the bottom half of the sprayer tight on the hose **4**. Remove it by snapping it off to the side with the point of a sturdy knife or a miniature screwdriver (don't cut yourself and don't lose the retaining ring) **5**. Remove the collar from the hose **6**. Then soak all the parts in hot vinegar for up to 1 hour to dissolve any lime on the parts or in the spray holes of the head. When you reassemble the parts, be sure the gasket ends up on top, not the spacer.

3 Pull the spacer (white) and gasket (black) off the sprayer hose tip.

4 A brass-colored U-shaped retaining ring holds the base nut on the hose.

HIGH-QUALITY ALTERNATES

All-metal higher-quality sprayers may be put together differently from their plastic counterparts. Typically, the spacer and seal will be replaced with an O-ring. Their basic disassembly and cleaning are the same as for plastic sprayers.

CLEANING A PULL-OUT SPRAY FAUCET

A pull-out spray faucet is a hand-held sprayer you can pull out from the main or auxiliary spout on the kitchen faucet. Indeed, in many sinks so equipped, it *is* the faucet spout.

This type of faucet head has evolved quite a bit over the years. Older models had a large heavy brass check valve in the line (to keep water from flowing backward) and these would chatter. If your pull-out chatters, replace it. Newer models feature a check valve made as a small plastic one-way button in the bottom of the handle.

All pull-out models work the same: Leave the head in place as a spout or pull it out and use its push-button as a sprayer. Like all sprayers, they sometimes have to be cleaned.

Start by pulling out the sprayer, unscrewing it from its hose, and setting it on the counter ❶. Using the countertop for leverage, unscrew the screen with an adjustable wrench, If there aren't flats for an adjustable wrench on the side of the screen, use a pliers. (Wrap electrical tape several times around the jaws to avoid damaging the aerator.) ❷ There are only two places to clean—the aerator screen and the check valve on the hose end. Check for debris in the screen and clean it out if necessary ❸.

➡ See "Cleaning Screens," p. 74.

The hose end of the sprayer should have a check valve, a small plastic spring-loaded button. Use a needle and pick to see if it pushes in and out without hanging up on sediment ❹. It should work fine as long as you have the debris cleared from it. If it needs cleaning, use a cotton swab.

1 Remove the sprayer from the hose and lay it down on the counter.

2 Unscrew the screen with an adjustable wrench or pliers.

3 Look at the reverse side of the screen and remove any debris.

4 Push on the check valve with a pick or miniature screwdriver to make sure it's clear and moving in and out smoothly.

CLEANING A SHOWER HEAD

Shower heads are one of the prime targets for lime buildup–calcium and magnesium sediments (the white stuff) that distill out of the water when it's heated in the water heater.

Some shower heads are throwaways. The bottom and top of fused heads, for example, will not come apart. If you have one of these, and it's clogged up, replace it. Higher-quality shower heads will offer some method for taking the unit apart, such as flats for an adjustable wrench. Even if the head is all plastic, it still may come apart.

Cleaning a shower head is a two-part procedure: dumping out the water-borne debris and then soaking away the iron stains and lime. In most cases, you will have to remove the shower head to clean it. Simply unscrew it from the chrome angled pipe coming through the wall.

Once removed, look down into the threaded collar for a screen that can be cleaned. Clean out any sediment with a pick or needle **A**. If that's not the problem, disassemble the shower head if you can. Unscrew the base nut with an adjustable wrench or pliers **B**. Lay the parts on a flat surface so none can get lost. Clean any parts with debris stuck to them. The only way to remove lime buildup from the tiny holes in the shower head is to dissolve it. Vinegar works fine. Heat the vinegar (do not boil) and pour it into a bowl. Put the shower end head down into the vinegar and leave for no more than 1 hour **C**. Soaking the head overnight will ruin the finish and metal. Do not add baking soda. Though the fizzing is impressive, the baking soda neutralizes the acidity that dissolves the lime.

Not all showerheads can be opened **up to be internally cleaned. The brass head (left) has a nut with flats that keeps its internal parts snug in the head. You can remove the nut and the parts. The low-cost plastic molded head (right) is fused together. It's designed as a throwaway, and you won't get this one apart.**

Even if the showerhead is all plastic, **you may be able to disassemble it. Many are designed to come apart so you can clean them.**

A Look for a screen and clean it.

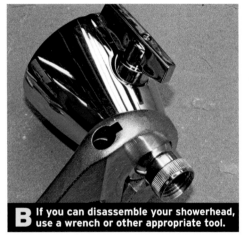

B If you can disassemble your showerhead, use a wrench or other appropriate tool.

C Invert the shower head and soak in hot vinegar for 20 minutes to 1 hour.

WHAT CAN GO WRONG

Vinegar can eat through the finish of the shower head and ruin it. Check your cleaning solution every few minutes.

CLEANING AN ULTRA-LOW-FLOW SHOWER HEAD

An ultra-low-flow shower head can save many gallons of water but can clog just as easily as a standard shower head.

Remove the head from the shower arm with a small pipe wrench or if the head has flats, use an adjustable wrench ❶. The inlet side has no screen to clog, but when you unscrew the front of the unit and pull the head off, you'll find an O-ring and disc inside ❷. A small pick can pull out the O-ring if it doesn't fall out; the disc will fall out right after. Do not lose them.

Clean out the holes if they have debris in them. A small pick or needle will crush any iron or lime clogging the holes ❸. Shake out any residue. To reassemble the unit, drop the disc into the shower head end, and then drop the little O-ring on top of that ❹. Twist the shower end clockwise on the base until it's snug. No tools are needed ❺.

1 Remove the ultra-low-flow shower head with a small pipe wrench or adjustable wrench.

2 The side facing the shower head has no screen to clog up. As you lift off the head, an O-ring and disc with holes will drop out.

3 Using a small pick or needle, clean out all the small holes.

4 Reassemble the unit by putting in the disc with the holes first, then the O-ring.

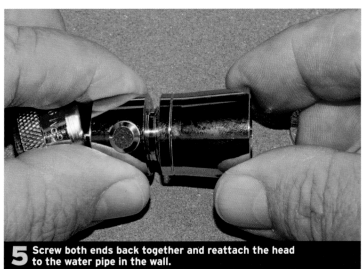

5 Screw both ends back together and reattach the head to the water pipe in the wall.

CLEANING A WASHER SCREEN

The clothes washer screen is perhaps the most consistently clogged of all the screens in appliances and spigots. Whenever you lose pressure and volume in either the cold or hot water lines in the clothes washer, always look to the screens. They are either in the washer itself (in the solenoid, accessed only from the back), where the hoses attach, or in the hoses themselves **A**, **B**. Use a small screwdriver or pick to pull out the screen **C**. Clean it as necessary or replace it. Never throw the screens away. Their purpose is to keep waterborne debris from jamming the solenoids. Debris can jam the solenoid closed or open—either way, it interferes with the operation of the machine.

It's best to have the screens in the hose collars at the hose bibb-hose interface because this is an easy place to get to. Leaving them on the washer solenoid means you have to pull the washer completely out to get to them because they are facing backward toward the wall. However, there is no requirement to leave them there. You can move them to the inlet end of the hose, the one in the washer box.

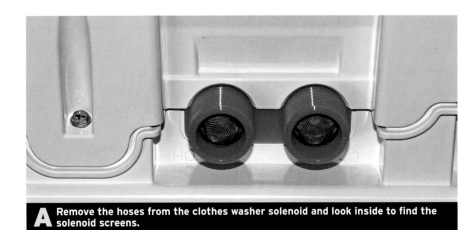

A Remove the hoses from the clothes washer solenoid and look inside to find the solenoid screens.

B Unscrew the supply hoses from the washer box.

C Use a small screwdriver to remove each screen; then clean them.

CLEANING A DISHWASHER INLET SCREEN

If your dishwasher takes forever to fill, a clogged inlet screen restricting the water flow might be the culprit. Open the lower access door at the bottom of the dishwasher and locate the inlet valve solenoid, typically on the bottom left. In some cases you may have to pull the washer out and put it on its back to have access to the solenoid **1**. Locate the solenoid valve, remove the inlet hose, and then unscrew the brass angle fitting. Pop the screen out with a pick or small screwdriver, and clean or replace it **2**. Reassemble the parts in the reverse order.

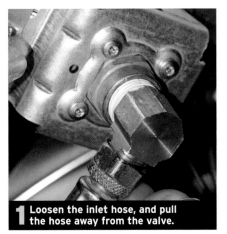

1 Loosen the inlet hose, and pull the hose away from the valve.

2 Pop the screen out, and clean or replace it.

CLEARING A SINK DRAIN

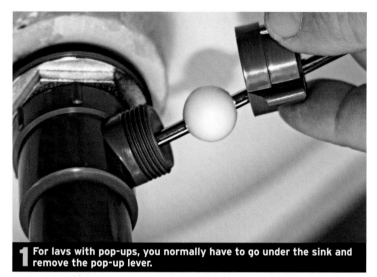

1 For lavs with pop-ups, you normally have to go under the sink and remove the pop-up lever.

2 Pull the pop-up up and out, and clean it if necessary.

3 With pop-up lever removed, run a small auger through the line.

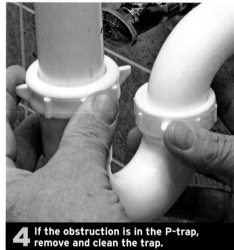

4 If the obstruction is in the P-trap, remove and clean the trap.

For sinks without a pop-up, you can send in a small hand auger to clear any stoppages.

The lav drain gets its fair share of clogs, mostly due to human hair and soap scum, and sometimes an occasional tooth-brush. Nine times out of 10, you can clear the lav drain by simply removing the pop-up from the center of the bowl and cleaning off the hair.

Some sinks, such as bowl sinks, do not have a pop-up. For these models you can often auger straight down into the vertical part of the trap. However, your cable may not be able to get through the trap itself. Be sure to check the drainpipes for leaks after you've tried any auguring.

To rod a sink with a pop-up you have to go under the sink and unscrew the cap holding the lever that activates the pop-up. Then pull the lever straight back and out. Do not lose any parts. Be wary of a washer that might come out as well. If it does, reinsert it with the concave part facing out toward the ball on the lever **❶**. Remove the pop-up. This is where most clogs occur. If the pop-up comes out clean, as the one shown here **❷**, you'll have to find the clog in another loca-tion. If it comes out with a mass of hair, clean it. With the pop-up removed, you can rod the line straight down from the sink **❸**.

You can also use a fish-hook grabber to pull out any debris and balls of hair. You may want to forgo using the auger and go straight to the P-trap. Put a bucket under the trap to catch the water that's in it. Then simply unscrew the two slip nuts and the trap will fall away **❹**. Clean the trap and make sure there is no blockage in the trap line going into the wall. If you still can't find the blockage, you may have to clean the trap line.

➡ See "Cleaning the Trap Line," p. 83.

CLEANING THE TRAP LINE

If you can't find the clog in the pop-out or the trap, you can remove the P-trap and P-trap arm as one unit and clean out the drain line in the wall.

Unscrew the slip nut from the tailpiece (the pipe that comes down from the sink drain). Then loosen the nut that holds the trap arm in the wall. Drop the trap from the tailpiece and work the trap arm out of the wall orifice ❶.

Many times the clog is at the elbow or T immediately behind the wall. Using a flashlight, look into the drainpipe that goes into the wall. Even if you can't see a clog, try inserting a bent coat hanger into the opening and give it a couple of twists. You might be able to grab accumulated hair with this tool and technique. You can also try a fish-hook remover (sold where any fishing supplies are sold) or rod the line with a small auger. Use any type of grabber mechanism you find practical to pull out the obstruction ❷,❸.

1 If desired, you can remove the complete P-trap assembly as one unit and get access into the drain line.

2 With the P-trap removed, you can rod or reach directly into the drain line to remove the obstruction.

3 Use any grabbing device practical to find the clog and remove it.

USING A DRAIN-CLEANING BLADDER

A practical device for removing clogs with water pressure has been a long time evolving, but drain bladders provide a welcome and useful tool. Just screw the bladder to a garden hose, insert it in the drainpipe, and turn the water on. The water will swell the bladder against the walls of the pipe, sealing it against the pipe and stopping the water from coming back and flooding the area. Then the bladder will send a jet of water streaming out the front to blow out the clog. The high-pressure water will also flush out any leftover debris in the line.

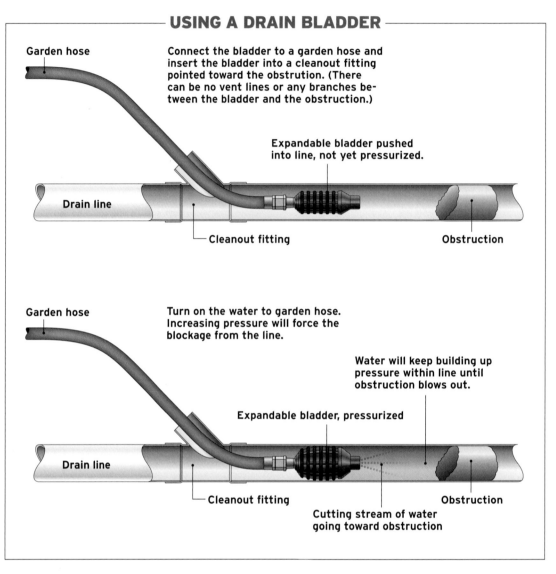

USING A DRAIN BLADDER

Garden hose

Connect the bladder to a garden hose and insert the bladder into a cleanout fitting pointed toward the obstruction. (There can be no vent lines or any branches between the bladder and the obstruction.)

Expandable bladder pushed into line, not yet pressurized.

Drain line

Cleanout fitting

Obstruction

Garden hose

Turn on the water to garden hose. Increasing pressure will force the blockage from the line.

Water will keep building up pressure within line until obstruction blows out.

Expandable bladder, pressurized

Drain line

Cleanout fitting

Cutting stream of water going toward obstruction

Obstruction

Screw the bladder onto a common water hose. When you turn the water on, the bladder will expand and seal the bowl.

Liquid drain cleaners

If liquid drain cleaners worked as well as advertised, we wouldn't need augers and other devices to clean drain lines. Chemical cleaners have their place for small clogs, but not for extensive blockages, such as one in an entire length of kitchen drain line.

Once you've used a chemical cleaner, you can't immediately rod the line. The acid will splash back on you. Wait at least 24 hours to break into a line with acid cleaner in it.

Wear safety goggles, a long-sleeve shirt, long pants, and rubber gloves. If you get splashed, run, don't walk, to any flowing water and bathe the area immediately. You have only seconds to get it off. If the cleaner bubbles up out of the drain or strainer, don't touch it. If you see fumes coming up, open the room to fresh air immediately. Do not breathe the fumes.

UNCLOGGING A SHOWER DRAIN

1 Try to clear the drain by sharply pushing and pulling a plunger.

2 If the plunger is unsuccessful, use an expanding bladder to dislodge blockage.

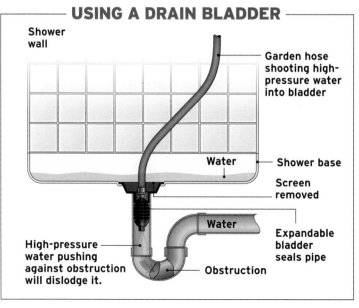

TRADE SECRET
At times it is easier to replace the entire trap or even the entire drain line rather than to spend hours clearing it.

PLUNGING A DRAIN

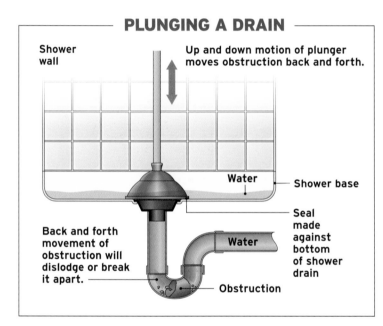

Shower wall

Up and down motion of plunger moves obstruction back and forth.

Water

Shower base

Seal made against bottom of shower drain

Back and forth movement of obstruction will dislodge or break it apart.

Water

Obstruction

CLEARING A SHOWER DRAIN WITH A HOSE

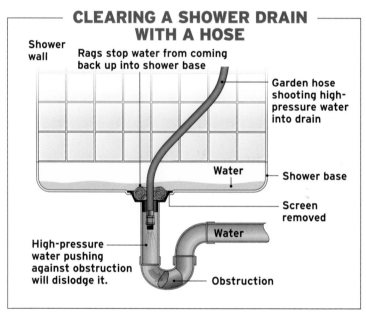

Shower wall

Rags stop water from coming back up into shower base

Garden hose shooting high-pressure water into drain

Water

Shower base

Screen removed

High-pressure water pushing against obstruction will dislodge it.

Water

Obstruction

USING A DRAIN BLADDER

Shower wall

Garden hose shooting high-pressure water into bladder

Water

Shower base

Screen removed

Water

High-pressure water pushing against obstruction will dislodge it.

Expandable bladder seals pipe

Obstruction

The shower drain can plug up with hair, oil, cleaners, and other sticky things. If you take the strainer off during use, as you should not do, the drain can collect larger objects, like toothbrushes and shaving gear. When unclogging a shower drain, first try a plunger ❶. Then go for the gold with a hose by itself or a hose and an expandable bladder ❷.

Plungers

The right plunger will make a perfect seal in the drain. When you press down and pull up the plunger, the end needs to be flexible enough to seal the perimeter of the drain. Otherwise you are wasting your time. In addition, the upper end has to be large enough to move a lot of water.

➤ See "Picking a Plunger," p. 13.

UNCLOGGING THE KITCHEN SINK DRAIN

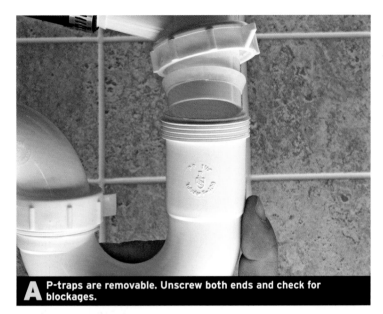

A P-traps are removable. Unscrew both ends and check for blockages.

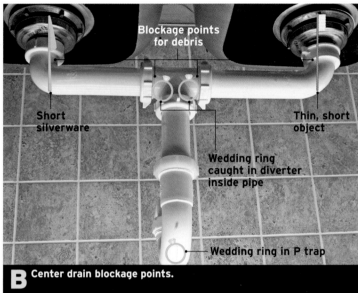

Blockage points for debris

Short silverware

Thin, short object

Wedding ring caught in diverter inside pipe

Wedding ring in P trap

B Center drain blockage points.

C Center drains have diverters at their Ts. These can cause blockage and catch rings.

D An option for center diverter drains is to go directly to the P-trap to check for a blockage.

Unclogging a two-bowl kitchen-sink drain presents unique problems. You often cannot plunge one side because the air will just push out the other drain. Besides even in a one-bowl sink, plunging rarely does any good because the stoppage needs to be removed, not just shoved away. Experience has shown that more than a few kitchen utensils can slide right through some bowl strainers.

The most common stoppages in a kitchen sink drain are silverware (knives) and toothbrushes. Such objects will usually get caught in the tailpipe below the strainer and of course, they have to be physically removed. Wedding rings, however, find their way to the trap. A wedding ring can also be in the drain diverter—inside the pipe where the water goes from horizontal to vertical.

Remove the trap or pipe and pour the water into a bucket **A**. Center drains are famous for getting blocked up at the center diverter due to the reduction of interior pipe diameter where it makes a 90 degree turn. Remove the T-fitting and check for blockage. You may have to disconnect one end from the strainer **B**,**C**. Or, you can go directly to the P-trap, hoping the problem is there, as opposed to in the T **D**.

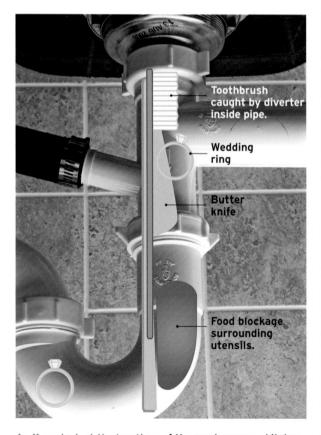

A plunger may not do much good in a two-bowl sink. There's often really no good way of keeping the plunged air from coming through the second drain.

Toothbrush caught by diverter inside pipe.

Wedding ring

Butter knife

Food blockage surrounding utensils.

An X ray look at the locations of the most common kitchen drain stoppages.

CLEARING A TUB DRAIN

Clearing a tub drain is not easy. Typically, the stoppage is in the trap, and to get to the trap, you have to cut out a finished ceiling from below. Sometimes the stoppage is in the elbow where the drain goes from horizontal to vertical under the tub. Again, getting to this is not easy. You can try rodding down from the overflow pipe in the tub, but that technique rarely works. Consider yourself lucky if the trap is in an open floor-joist space in the basement. Sometimes it's easier to replace the trap than to drill out the hardened soap and other granular substances. In the end, if you can't remove the blockage, you'll have to remove the trap.

OPENING UP A TUB DRAIN

Overflow cover

Overflow pipe

Leave drain alone, you cannot rod from here.

Possible restriction site

Full-sized P-trap

1. Attempt to rod blockage through the tub overflow pipe. Remove overflow cover and linkage. If this doesn't work because linkage or restrictions in the overflow pipe block your rodding equipment, go to step 2.

2. Remove trap from under floor and clean it.

Blockage is almost always in the trap, due to soap granules and hair sticking to oily film on wall of pipe. This hardens like concrete.

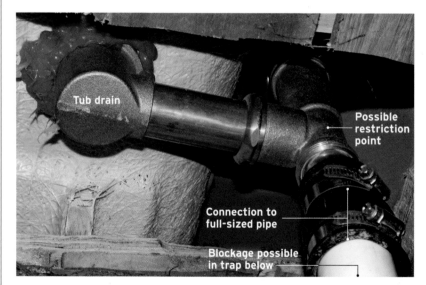

Tub drain

Possible restriction point

Connection to full-sized pipe

Blockage possible in trap below

You may have to work below the tub and remove the trap to clear the blockage. This may mean removing the finished ceiling in the room below the tub.

CLEARING A STOPPED-UP TOILET

1 The first line of attack is with a plumber's helper.

2 If the plunger doesn't work, try using a hand-cranked auger.

3 An expandable bladder can seal off the opening and send a high-pressure stream of water into the waterway.

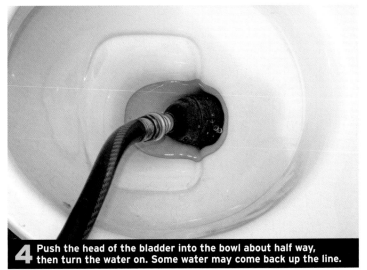

4 Push the head of the bladder into the bowl about half way, then turn the water on. Some water may come back up the line.

Occasionally, toilets stop up, and their reasons for doing so often depend on their age. Early model 1.6 gpm toilets may need to be replaced. Old toilets sometimes stop up because the waterway was made with a very rough unfinished porcelain. This roughness grabs at the toilet paper, slowing the flush. Compounding the problem are mineral deposits. If your toilet is old, consider replacing it. Most of today's toilets have a fine, smooth porcelain in their waterways, but they still stop up, especially when toys, toothbrushes, and other objects are dropped into them.

The first step in clearing a toilet is the old standby, the "plumbers helper." **❶**

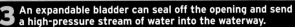

➤ See "Picking a Plunger," p. 13.

If that doesn't work, you have to go to a hand-crank auger **❷**. The auger slides along the waterway, and as it turns (because you crank it), it can catch and dislodge an obstruction. However, if the toy, toothbrush, or other obstacle is turning sideways with the auger head, the auger won't help. You know this is happening when the toilet flushes once well, and then it stops up again.

➤ See "How an auger works," p. 89.

Your third line of defense is an expandable bladder **❸,❹**. Assuming you can get a hose into the house, try using a 1½-in. expandable bladder. It will seal off (almost) the entrance and send a high-pressure stream of water into the waterway. Some of the water may come back, so you will have to watch the bowl so it doesn't overflow.

>> >> >>

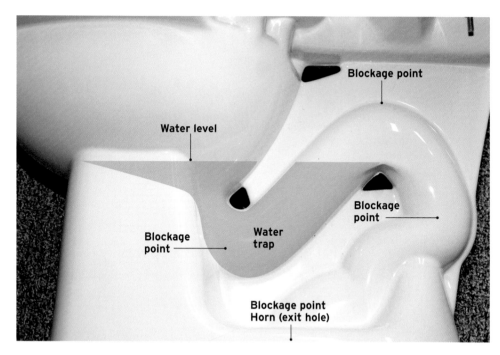

Water level

Blockage point

Blockage point

Water trap

Blockage point

Blockage point
Horn (exit hole)

The toilet waterway undulates like a snake. The blockage points are always at the bends.

How an auger works

To use an auger, first withdraw all the cable into its plastic body. Then insert the auger end into the bowl, pushing the auger so it follows the waterway within the toilet. To aid the insertion, turn the handle. If you feel an obstruction, keep turning the handle so the auger end can tear at it. Eventually, the obstruction will rip apart. An auger will normally work on common stoppages. However, if the stoppage is an object, such as a toy or toothbrush, it may turn and let the auger head through. Then an auger won't help. You know this is happening when the toilet flushes well with water only but will stop up when toilet paper is thrown in (it catches on the toy).

X-RAY VIEW INTO AUGERING THE TOILET

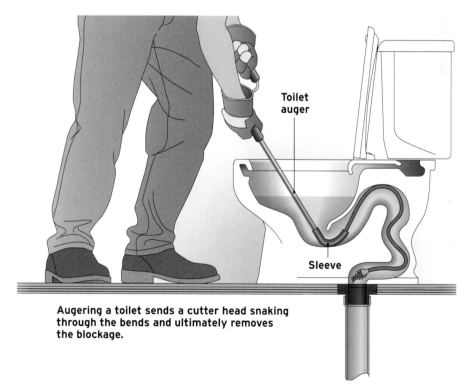

Toilet auger

Sleeve

Augering a toilet sends a cutter head snaking through the bends and ultimately removes the blockage.

CLEARING A STOPPED-UP TOILET (CONTINUED)

5 A bolt grabber opens its jaws when you push a plunger and closes when you release the plunger.

6 A bent coat hanger can sometimes hook an obstruction.

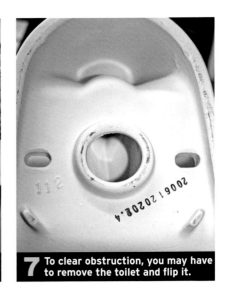

7 To clear obstruction, you may have to remove the toilet and flip it.

10 A bolt grabber has a longer reach than a fish hook remover, and can sometimes hook an object not visible from the horn.

11 Now we are getting serious. Take the toilet outside. Use the hand-crank auger to rod the waterway backward from the horn.

WARNING

An existing crack (some are hairline cracks) on a toilet can mean it will break apart as it is being pulled. Be careful: Broken porcelain is razor sharp. Be sure to wear heavy gloves.

Your last line of defense will be a bolt grabber **5** or a bent coat hanger **6**. Each can slide along the curves of the waterway and sometimes grab the obstruction. A bolt grabber (available at automotive stores) has a claw that opens and closes as you push and let go of a spring-loaded plunger on the other end. And sometimes the old method of using a coat hanger might work.

If nothing works, you have to get serious. You have to pull the toilet and look in from the bottom (the horn) to try to locate and remove the stoppage **7**. This means draining the toilet, unhooking the water line, pulling the toilet, and flipping it without breaking it.

➤ See "Draining a Toilet Tank and Bowl," p. 113.

Once flipped, look into the horn to see if the obstruction (typically a toy) can be seen. If you are lucky, it will be right at the horn. A fish hook grabber (sold at discount, hardware, and fishing stores) has an openable toothed jaw that can reach in and grab the obstruction **8**,**9**. You can also reach

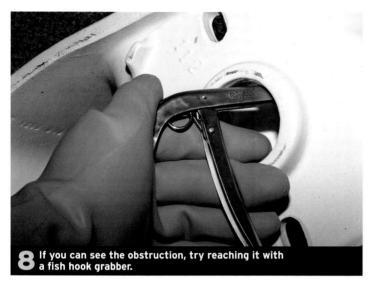

8 If you can see the obstruction, try reaching it with a fish hook grabber.

9 Insert the fish-hook remover carefully and try to maneuver its jaws so they will close around one edge of the obstruction.

12 If possible, insert a water hose into the bowl and turn it on while augering from the horn.

TRADE SECRET

If the toilet is one of the first 1.6-gal. flushes that came onto the market, and it constantly has trouble flushing, it should be replaced. Any money spent on it is wasted.

in with the bolt grabber or coat hanger **10**. Finally, carry the toilet outside.

➤ See "Replacing a Toilet," p. 132.

Insert the hand-crank auger into the horn, and try to dislodge the obstruction **11**. If possible, as you pull back on the auger, insert a hose into the hole at the front end (the toilet bowl) so water pressure will also be shoving on the obstruction as you pull the auger back **12**. If none of this works, then you have no choice but to buy a new toilet.

RODDING A DRAIN LINE

1 To rod a drain line, you have to find and open a cleanout closest to the clog.

2 Very carefully and slowly open a small crack to see if there is any fluid underneath. If none, remove the plug.

3 Insert a sewer tape into the line.

or

A rodding machine will send a cutter head into the line. A foot switch (inset) is mandatory for safe operation.

4 With blockage cut through, run a hose in the line full force for several minutes.

RODDING A BLOCKED MAIN DRAIN

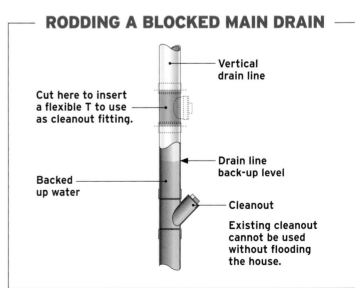

Vertical drain line

Cut here to insert a flexible T to use as cleanout fitting.

Drain line back-up level

Backed up water

Cleanout

Existing cleanout cannot be used without flooding the house.

FINDING THE BLOCKAGE

If the cleanout is below the surface of the water backup, all of whatever is in the lines will flow out and onto the floor when you open the cleanout. This is why one should always have a cleanout outside of the building. To tell where the backup level is, fill the line and tap it in several places. A dull thud means the line is full. A hollow sound means it is empty.

Rap here and it should sound hollow

Feels warm on right side and cold on left—the blockage is in middle.

Rap here, and it should sound solid because it is full of water.

Drain or sewer line

Blockage

If you let hot water flow down pipe until it backs up in pipe, it will sometimes feel warm to touch above blockage.

Using a sewer tape or machine to cut through the debris inside the drain lines is called rodding the lines. A sewer tape will not cut roots, so for outside lines, use a machine rated for that purpose. Roots are one of the main sources of blockage in outdoor lines. Never plant a tree over a drain line.

Make sure you wear old clothes you can throw away as well as safety glasses and gloves. You also can try to rod the main line from the stack on the roof, but that is quite dangerous. To rod a cleanout, use a large pipe wrench and turn the cleanout plug counterclockwise ❶. Get it loose, keep pressure against it, but do not remove it. If there is drain water backed up behind it, it will start to seep out. If that happens, rethread the plug back into the cleanout. You will then have to find a way into the pipes at a higher level or wait a day for the fluid to go down. If there is no fluid seepage, remove the plug ❷. Insert a sewer tape into the line (be sure to wear gloves) pointing downward into the blockage. Push the tape forward into the pipe, rotating the reel clockwise as you unroll it. When the progress of the tape requires more forward pressure, back

the tape off slightly and push it forward again. What you're attempting to do is force the tape to cut through the blockage. Use hard back and forth jerking actions. Or, you can rent a rodding or sewer-line cleaning machine. It will also make its way through the lines more easily, but you will still have to push it. Since you'll need both hands to push the cable into the sewer line, you'll control the power with a foot switch. If the cable reaches a snag (at a turn in the lines or when first encountering a blockage, for example) it can rapidly twist itself around your hands with enough force to break bones. At the first sign of twisting, immediately release the foot switch to cut the power to the machine, flip the directional switch, back off the line with the power in reverse, and then start the line forward again ❸. Once the sewer tape or machine has cut through the blockage it is mandatory to send high pressure water through the lines to carry the debris away ❹. Be sure to get clear instructions from your rental store staff and be careful.

REPAIRING FAUCETS

FAUCETS, REGARDLESS OF manufacturer, have but one function: to control the water supply. However, they all do it differently—some with seals and springs, some with cartridges and seals, some with cartridges and O-rings. Then there's an array of handle designs to make all the parts do what they're supposed to do. There are one-handled units, two-handled models, some with three handles, some with handles that turn, and others that are pulled and pushed.

Adding to this chaos are the mechanics of some of today's faucets. Manufacturers' current trends abandon the simple three-piece design of the past (handle, stem, body) and replace it with complicated designs that are hard to take apart, harder to put back together, and are costly. Add in labor costs, and the total bill for a repair person to fix your faucet can exceed its original cost. Today, more than ever, it pays to both install and fix a faucet yourself.

TIPS FOR FAUCET REPAIR

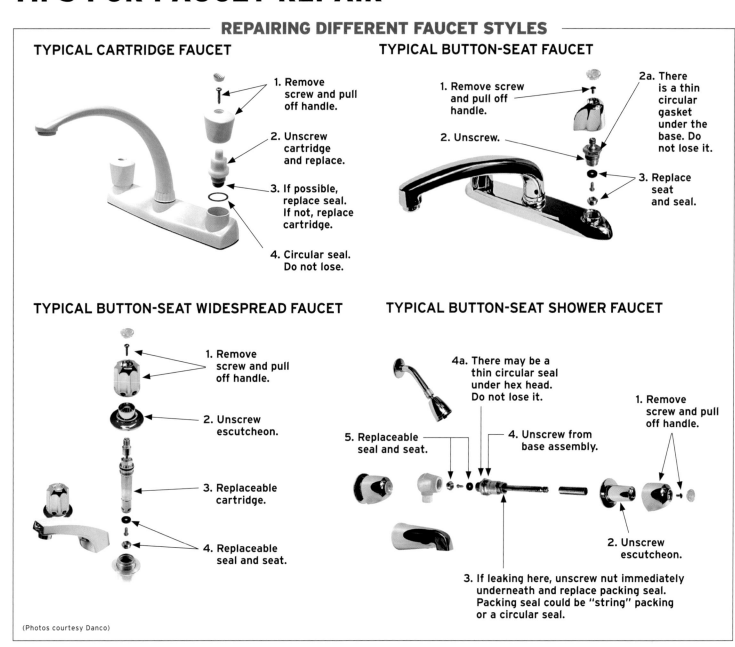

REPAIRING DIFFERENT FAUCET STYLES

TYPICAL CARTRIDGE FAUCET

1. Remove screw and pull off handle.

2. Unscrew cartridge and replace.

3. If possible, replace seal. If not, replace cartridge.

4. Circular seal. Do not lose.

TYPICAL BUTTON-SEAT FAUCET

1. Remove screw and pull off handle.

2. Unscrew.

2a. There is a thin circular gasket under the base. Do not lose it.

3. Replace seat and seal.

TYPICAL BUTTON-SEAT WIDESPREAD FAUCET

1. Remove screw and pull off handle.

2. Unscrew escutcheon.

3. Replaceable cartridge.

4. Replaceable seal and seat.

TYPICAL BUTTON-SEAT SHOWER FAUCET

4a. There may be a thin circular seal under hex head. Do not lose it.

5. Replaceable seal and seat.

4. Unscrew from base assembly.

1. Remove screw and pull off handle.

2. Unscrew escutcheon.

3. If leaking here, unscrew nut immediately underneath and replace packing seal. Packing seal could be "string" packing or a circular seal.

(Photos courtesy Danco)

No matter how complicated a faucet looks in an illustration on the manufacturer's spec sheet, repairing a faucet follows several general steps that apply to all. First you remove the cap and whatever kind of screw holds the handle on. Then comes some kind of retainer securing the valve parts inside the faucet body. After that and maybe some trim pieces, there is some kind of sealing device—springs and seals, O-rings, or button seals that press against a seating surface to regulate the water flow. The illustrations above will give you an idea how similar the process is for a few of the most common faucets.

TRADE SECRET

You paid a lot of money for the fancy finished surface even if it's chrome, so take care when cleaning. Any white spots on the faucet are minerals from the water; they act like sandpaper on the finish. Thus you want to lift them off, not rub them in. To keep the finish shining, use a very mild soap that doesn't leave a residue. Even better, use a lens cloth to lift off the minerals. These cloths are sometimes free at outlets for eyeglasses.

FAUCET REPAIR TOOLS

Moen's cartridge puller can remove **cartridges you can't get out by other means.**

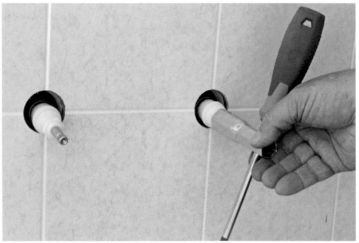

Special large-diameter sockets are used for unscrewing stems from some 2-and 3-handled tub faucets.

Use a stepped hex-head wrench **when removing and replacing faucet seats. Use grinders to smooth a valve seat you don't want to replace.**

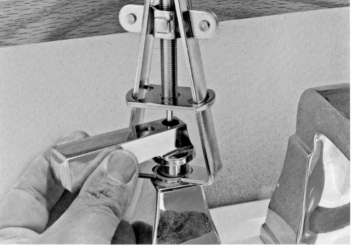

For stuck handles, it's either use this **handle puller or toss the faucet.**

Although you can remove most cartridges with a pliers and take out springs and seals with a pencil, there are commercial tools made for maintenance problems household tools won't fix.

In addition to the more common tools shown here, you may also need small Allen (hex-head) wrenches or Torx wrenches to loosen the set screws on faucet handles.

➤ See "Specialty Tools," pp. 6-7.

Don't lose the parts. There are few less frustrating things than losing a faucet part down the drain. To prevent this, pull the drain arm up to lower the seal in the drain. If the drain system doesn't work, place a cloth over the hole. Then as you remove any parts, put them in a container on the sink next to your work. Remember, if you do lose a part down the drain, its not gone forever. It's in the P-trap, and you can retrieve it by removing the trap.

REPAIRING KITCHEN CARTRIDGE FAUCETS

Repairing cartridge faucets with springs and seals or O-rings is a straightforward project; in 15 minutes at the most you'll have the faucet torn down and reassembled with new parts, stopping that pesky (and expensive) leak. This doesn't count the trip you'll make to the hardware store, of course. Like all faucets, the parts made by different manufacturers vary widely. If you don't know the manufacturer and model number of your faucet, disassemble it as shown here, and take it to your retailer for an exact replacement.

Repairing a two-handled cartridge faucet with springs and seals

Cartridge faucets using a spring-and-seal system work quite well and can last a long time. But every once in a while, you have to either stretch or replace the springs and/or replace the seals. As with most faucets, you start by popping off the cap to remove the handle ❶, and withdrawing the handle screw with a Phillips® screwdriver ❷. Once the handle is off, remove the decorative skirts ❸. On some models, these unscrew; on others, they just pull off. After you take the skirt off, unscrew the cartridge hold-down nut and set it aside ❹. Then pull out the cartridge ❺. Looking down into the faucet body, you'll see the seals (the springs are under them). Gently insert a dull pencil into the opening of the seal and lever both the seal and spring out of the body at a slight angle ❻. Be careful you don't break the pencil lead off in the process or you'll have to flush out the broken piece before putting things back together. You can also use a large paper clip with a small angle bent in the end.

If the faces of the seals don't look scarred, you might get by for a while by stretching the springs. Just grab both ends of a spring and pull gently. Reassemble the parts in the reverse order, but don't count on this solution to last forever. If stretching the springs doesn't stop the leak, replace both seals and springs with exact replacements.

1 Pop the cap off the handle with the tip of a utility knife or miniature screwdriver.

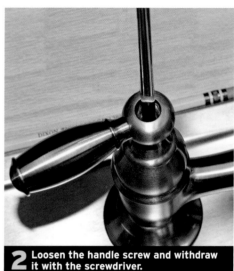
2 Loosen the handle screw and withdraw it with the screwdriver.

3 Pull or unscrew the decorative skirt and set it aside.

4 Using an adjustable wrench, remove the cartridge hold-down nut.

5 Pull out the cartridge.

6 Using a pencil, pull out the springs and seals and replace them.

Repairing a two-handled cartridge faucet with O-rings

After springs and seals, O-rings are the next most common mechanism for controlling water inside a faucet. They're easy to repair, and the O-rings are available everywhere. To start, slide the edge of a knife blade between the cap and the handle and pry up ❶. Set the cap aside and do not lose it. Using a Phillips screwdriver, remove the handle screw ❷. Then lift off the handle and decorative skirt and remove the cartridge retaining nut, setting it aside ❸. Pull the cartridge out of the faucet body using pliers ❹. Replace the cartridge or the two O-rings—either will stop the dripping. To replace the O-rings, pry them out of their slots and roll the new ones over the cartridge body and into the grooves, taking care not to nick the cartridge body. This type of cartridge features a molded tab which fits into a keyed slot in the cartridge housing, keeping the cartridge in the correct position. When reinstalling it, make sure to rotate the cartridge so the tab drops into the slot ❺.

If the spout leaks around its base, you'll need to remove it. Unscrew the skirt, and pull the spout up and out. Replace the O-ring(s) and reassemble.

1 Using a thin blade, remove the cap from the handle.

2 Unscrew the handle nut, if any, and withdraw the handle screw.

3 Remove the cartridge cover nut and set it aside.

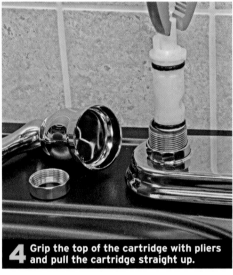

4 Grip the top of the cartridge with pliers and pull the cartridge straight up.

Slot for cartridge key

5 When reassembling, make sure the keyed tab on the cartridge fits into the slot.

If the spout leaks around its base, **unscrew the skirt and lift it off. Replace the O-rings.**

REPAIRING A BALL-STYLE FAUCET

If you learn how to repair one single-handle Delta® or Delta copy-cat faucet, you can repair them all. Regardless of their use, all employ a rotating ball with ports that allow water to flow. Pulling the handle forward moves a solid face of the ball over the seals in the base, shutting off the water.

Make sure you close the pop-up drain to keep small parts from falling into the drain and keep a bin handy for corralling the parts as you remove them.

➡ See "Don't lose the parts," p. 97.

Use an Allen (hex-head) wrench to loosen the handle set screw, then pull the handle off ❶. If the cap has faces for a wrench, remove the cap with an adjustable wrench. Use slip-joint pliers or a pipe wrench on a cap with a knurled edge ❷. Pull the stainless steel ball up and out of the faucet body ❸. Using the short end of your Allen wrench (or a bent paper clip), pull out the springs and seals, noting whether they're tapered or not. Replace the springs and seals with new ones, and make sure that tapered seals go in the same way as the originals ❹. When you're putting things back together, set the ball in so the slot in the ball rides over the pin in the valve body ❺. Align the key in the plastic cover so it fits into the slot on the faucet body. Move the ball around to make sure the ball shaft will seat in the front of the V on top of the plastic cover ❻. Then install the cap, tightening it so it both moves freely and turns the water completely off.

1 Using an Allen wrench, loosen the set screw and remove the handle.

2 Remove the cap with an adjustable wrench.

DELTA BALL FAUCETS

The internal design of all single-handle ball-and-spring faucets is the same for kitchen, bath, and shower installations, regardless of the type of handle the faucet is fitted with. The illustrations below apply to all ball-and-spring faucets.

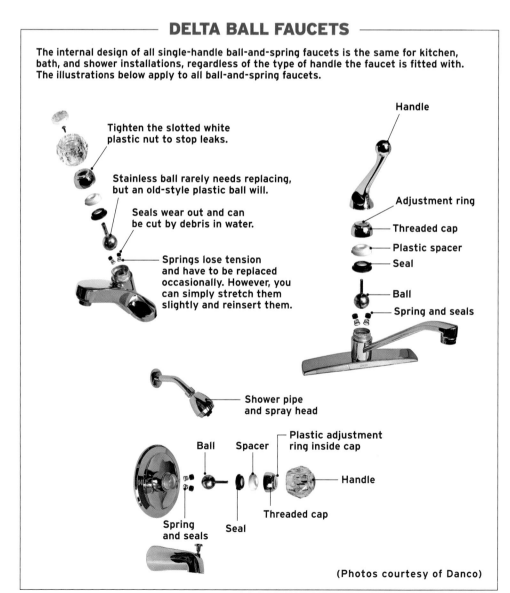

Tighten the slotted white plastic nut to stop leaks.

Stainless ball rarely needs replacing, but an old-style plastic ball will.

Seals wear out and can be cut by debris in water.

Springs lose tension and have to be replaced occasionally. However, you can simply stretch them slightly and reinsert them.

Handle

Adjustment ring

Threaded cap

Plastic spacer

Seal

Ball

Spring and seals

Shower pipe and spray head

Ball

Spacer

Plastic adjustment ring inside cap

Handle

Threaded cap

Spring and seals

Seal

(Photos courtesy of Danco)

3 Withdraw the ball to expose the seals and springs underneath.

4 Using a small Allen wrench, pull out the springs and seals and then replace them.

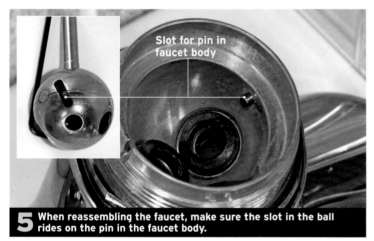

Slot for pin in faucet body

5 When reassembling the faucet, make sure the slot in the ball rides on the pin in the faucet body.

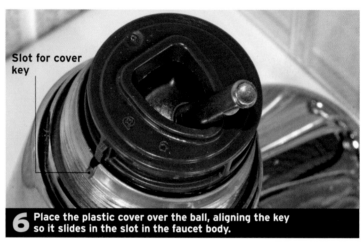

Slot for cover key

6 Place the plastic cover over the ball, aligning the key so it slides in the slot in the faucet body.

Quick fix Before you embark on a complete tear-down of a ball faucet, you can take a stop-gap measure. At the first sign of a drip, remove the handle and tighten the cap slightly. That puts pressure on the ball, which in turn puts pressure on the seals and stops the drip. You can keep doing this until the handle won't move. Then it's time for an overhaul. Of course, if the seal is scratched or the ball is worn, no tightening will stop the drip. Those problems call for replacements.

You'll find two types of springs and seals for ball faucets, one for older faucets and one for newer models. Make sure you use the correct replacements.

REPAIRING A PRICE PFISTER CARTRIDGE FAUCET

1 From the rear, loosen the handle set screw with an Allen wrench.

2 Pull the handle off the stem and set it aside.

The Price Pfister cartridge faucet system is quite simple, and because of that, it's pretty reliable. The O-ring seals at the bottom of the cartridge take the most wear. If you have a new set on hand before you tear your faucet down, you're ready to go. But if you don't mind the faucet being down for a few hours, disassemble it and take the cartridge to a hardware store and get new O-rings or a replacement cartridge. If parts are not available locally, contact Price Pfister on the Internet at www.pricepfister.com.

A Price Pfister faucet handle is held in place with a set screw. Using an Allen wrench, loosen the screw to release the handle ❶. Remove the handle and the cover plate and use a small pipe wrench to remove the cartridge retainer ❷,❸,❹. Remove the cartridge by hand or by jerking it straight up with a pliers ❺. Flip the cartridge over and you'll see the replaceable O-rings. Pull out the O-rings and wipe off any debris on the bottom of the cartridge ❻. Clean out the holes if necessary. Make sure there aren't any cuts or scratches across the holes. Push new O-rings in the holes, making sure they're properly seated. Then slip the cartridge into the faucet body, making sure that the tab on the cartridge fits into the keyed slot in the faucet body ❼. Replace the retainer, cover plate, and handle.

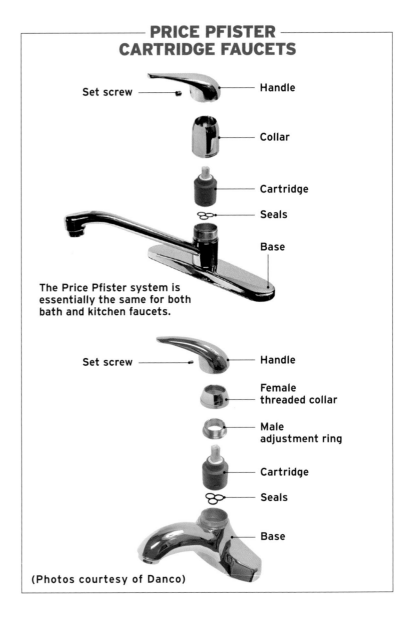

PRICE PFISTER CARTRIDGE FAUCETS

Set screw — Handle
— Collar
— Cartridge
— Seals
— Base

The Price Pfister system is essentially the same for both bath and kitchen faucets.

Set screw — Handle
— Female threaded collar
— Male adjustment ring
— Cartridge
— Seals
— Base

(Photos courtesy of Danco)

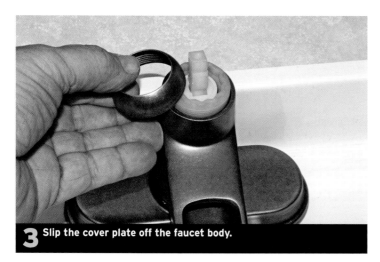

3 Slip the cover plate off the faucet body.

4 Remove the retaining nut with a small pipe wrench.

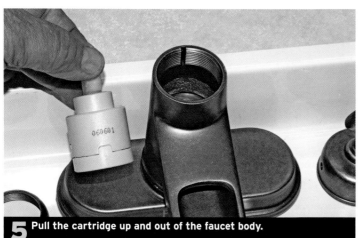

5 Pull the cartridge up and out of the faucet body.

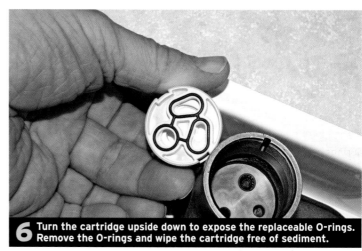

6 Turn the cartridge upside down to expose the replaceable O-rings. Remove the O-rings and wipe the cartridge free of sediment.

7 Replace the cartridge with its key in the slot of the faucet body. Reassemble.

Repairing almost any cartridge faucet requires the same steps—remove the handle and retaining nut and pull out the cartridge.

REPAIRING A NEW-STYLE MOEN CARTRIDGE FAUCET

The Moen cartridge system has evolved substantially over the years. Early designs lasted a long time (thousands, if not millions, are still being used). Their design is uncomplicated, they repair easily, and parts are still available. The newer Moen design, shown here, uses more parts and is far more complicated to take apart and put back together. Nevertheless, these faucets are not impossible to repair.

The new style uses a Torx screw to secure the handle. Look for it on the front collar of the handle. Loosen the screw and remove the handle ❶. Next to go is the handle support, a piece new to the faucet design used as an interface between the handle and the body. Loosen its retainer screw, pull out the retainer, and set both in your parts box ❷. Inside the faucet body is a threaded circular housing; unscrew it and set it aside ❸. Then unscrew and remove the trim collar ❹. Using a small-diameter machine screw or one side of needle-nose pliers, lever out the U-clip ❺,❻. Pull out the cartridge holder ❼. Now comes the cartridge. Grab the cartridge stem tight with pliers and jerk up ❽. Some cartridges may be hard to remove due to corrosion or sediment buildup, but if you jerk hard enough you'll get it out. Replace the cartridge and reassemble the faucet in the reverse order.

MOEN CARTRIDGE FAUCET

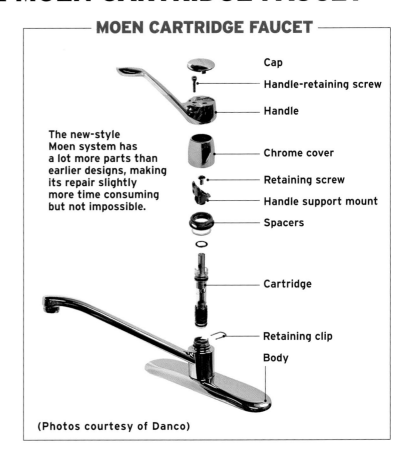

The new-style Moen system has a lot more parts than earlier designs, making its repair slightly more time consuming but not impossible.

- Cap
- Handle-retaining screw
- Handle
- Chrome cover
- Retaining screw
- Handle support mount
- Spacers
- Cartridge
- Retaining clip
- Body

(Photos courtesy of Danco)

GETTING THE RIGHT CARTRIDGE

Because there are several cartridge designs, you'll generally be better off if you disassemble your faucet and take the old cartridge to your hardware store. That way you can make sure you get an exact replacement. Some retailers will stock cartridges in both brass and plastic. Just match the replacement to the one you have.

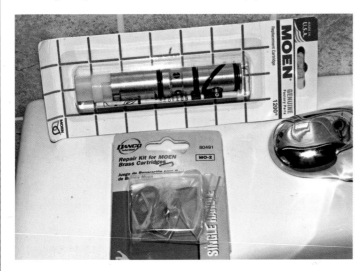

1 Loosen the Torx set screw and remove the faucet handle.

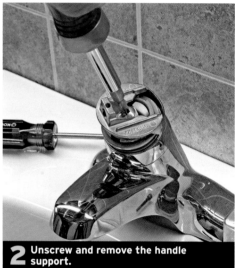

2 Unscrew and remove the handle support.

3 Unscrew the inner housing and pull it out of the faucet body.

4 Remove the chrome trim collar and set it aside.

5 Insert a small screw, miniature screwdriver, or needle-nose pliers in the U-clip hole and lever the clip from the slot.

6 Remove the U clip and set it aside.

7 Remove the cartridge retainer.

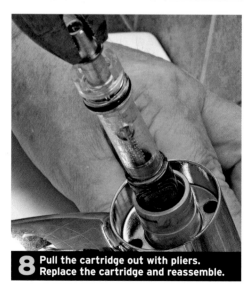

8 Pull the cartridge out with pliers. Replace the cartridge and reassemble.

REPAIRING A MOEN PUSH-PULL CARTRIDGE FAUCET

Although older-style Moen faucets featured an uncomplicated design, they would exhibit one primary problem. Hard-water mineral deposits or lubrication loss would make the cartridge hard to pull up and push down. Nevertheless, the older faucets are easy to work on. The procedure for shower and sink faucets is essentially the same.

Unless you have the original paperwork with the cartridge number, you will have to disassemble the faucet, and take the cartridge to your hardware store to find an exact match. If there's no lube on the replacement cartridge, and none is provided or for sale, be sure to lube it with plumber's grease before you reassemble the faucet.

You start out, as with any other faucet, by removing the handle. If you can't get the handle off because the set screw is stripped out, you'll have to drill out the set screw or break off the handle,

remove the handle, and purchase a replacement **❶**. Once the handle is off, remove the two long escutcheon screws and pull off the plate **❷**. At this stage, if you're replacing the entire valve body, you'll have to work on the back side of the shower wall, but if you're just replacing the cartridge, continue with the disassembly.

A stainless cartridge collar will now be evident—it has an indent that must always point up. Remove the collar, then using needle-nose pliers, grab the top of the U-clip and pull up. The U-link will come straight up and out **❸**, **❹**. Next, grab the cartridge stem with a pliers and pull the cartridge straight out **❺**. Sometimes the cartridge will stick, but pull hard enough and it will come. Wipe the inside of the cartridge housing with a rag to clear it from sediment. Lubricate and replace the cartridge, then reassemble the faucet **❻**, **❼**. Just remember to install the collar with the indent up, toward the shower head or faucet spout.

MOEN CARTRIDGE FAUCETS

The old-style Moen system lasts a long time and is simple to repair. Look for cartridges at all hardware stores.

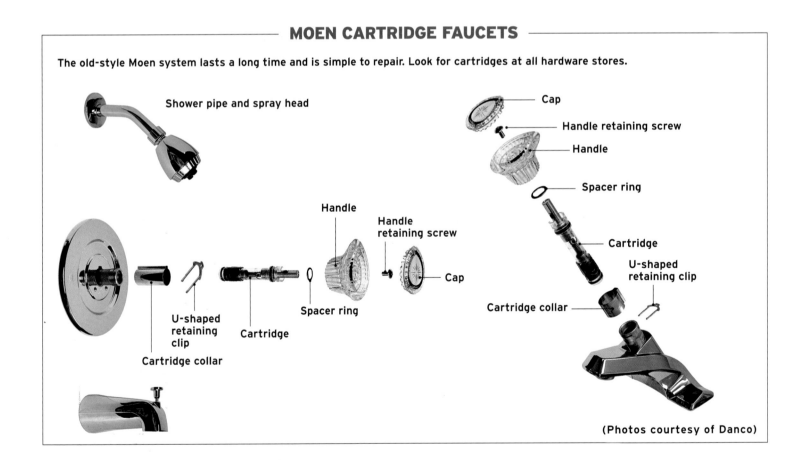

Shower pipe and spray head

Handle

Handle retaining screw

Cap

Spacer ring

U-shaped retaining clip

Cartridge

Cartridge collar

Cap

Handle retaining screw

Handle

Spacer ring

Cartridge

U-shaped retaining clip

Cartridge collar

(Photos courtesy of Danco)

1 Remove the handle, breaking it off if its set screw is stripped out.

2 Remove the escutcheon plate screws and pull off the plate.

3 Pull the cartridge collar forward to remove it.

4 Then use a pliers to pull the U-shaped retaining clip straight up and out

5 Grab the stem of the cartridge with a pliers and pull the cartridge with a firm jerk. Replace the cartridge. Remember to grease the new cartridge.

6 Attach the escutcheon plate.

7 Then reattach the old handle or a replacement.

REPAIRING A TWO-HANDLED TUB/SHOWER FAUCET

A typical two-handled spring-and-seal-type tub/shower faucet will have a chrome tube covering the valve assembly with threads at the back and a squared end up front. Inside, you'll find a cartridge and a seal and spring. If your faucet drips, you need new springs and seals. Don't replace just one. If only one leaks now, the other will leak soon. It's better to replace both while you've got your tools out. To repair the faucet, remove the chrome cover with an adjustable wrench slipped over the squared front end of the chrome tube ❶. With the protective tubes removed, the cartridges will usually pull straight out. If they stick, hold a pliers tight on the cartridge stem and pull hard ❷. Then using a pencil or other small-tipped tool, remove both the spring and the seal ❸. Replace the spring and seal, and reassemble the faucet.

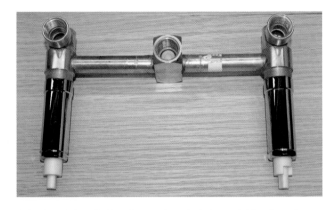

A typical 2-handled **spring-and-seal-style cartridge faucet.**

1 Loosen the protective chrome cover with an adjustable wrench.

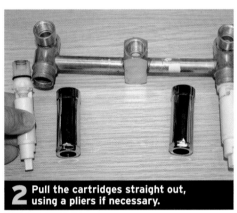

2 Pull the cartridges straight out, using a pliers if necessary.

3 Pry out both the spring and seal. Replace seals and springs on both valves.

REPAIRING A LEAKY HOSE BIBB

Hose bibbs leak at the stem nut and through the spout. Before you start unscrewing the nut, however, turn the handle full-on ❶. Loosen the packing nut with an adjustable wrench, then with your fingers ❷.

Then unscrew the stem and pull it from the body. If there's "packing," a leaded string wrapped around the shaft, replace it with a plastic packing disc. If the black seal is cut or damaged or worn, remove the center screw

and the seal ❸. Take the old one to the hardware store to get an exact replacement. Attach the new seal, set the stem back in the body, turn the handle down fully, and tighten the packing nut with a wrench.

Potential leak

Packing nut

Leaky internal seals here will cause drips here

1 Back the handle off all the way before loosening the packing nut.

2 Loosen the packing nut, unscrew it, and unscrew the stem from the body.

Button seal

3 Check the seal for worn spots and replace.

REPAIRING A LEAKY BOILER DRAIN

A boiler drain is hard to tear apart. The handle and packing nut will come away easily, but removing the stem from the body takes considerable muscle. You have to remove the drain from the boiler and clamp it in a heavy-duty vise that is securely bolted down. To further complicate repairs, you can't use a pipe wrench or an adjustable wrench to remove the stem because the flats are tapered. Only a deep-well socket and long breaker bar will get the stem off. Replacing a boiler drain is much easier than repairing one. But if none is available, you have to fix what you have.

Start by draining the water from the unit using a garden hose. Then disconnect the hose and remove the drain spigot from the boiler. Lock the drain in a vise. Remove the handle nut and handle ❶. Then unscrew the packing nut ❷. Inside the nut is the packing washer or stem seal. Replace this washer if the faucet leaks around the handle stem ❸. Now to the hard part. Use a deep-well socket of the exact size to loosen the stem body. Some stem nuts are standard, others are metric. Make sure your have the right socket by slipping on several sizes, selecting the one with the tightest fit. Attach a breaker bar to the ratchet and apply as much counterclockwise torque as you can. It will take a lot of muscle but the stem will come loose ❹. Pull it free and check the seal for cuts and abrasion ❺. Remove the seal-holding screw and replace the seal screw. Reassemble the unit, replace it on the boiler, and refill the boiler, following the manufacturer's instructions for any required bleeding of the system.

1 Remove the handle nut and the handle with pliers or small box wrench.

2 Remove the packing nut. Inside is the stem seal.

3 Replace the stem seal to stop a stem leak.

4 Using a deep-well socket and lots of muscle, unscrew the stem.

5 Check the seal for cuts and wear. Replace to stop any spout drips.

REPAIRING TOILETS

THE TOILET IS A COMPLEX DEVICE, made more so by the constant appearance of new designs and the reevaluation and redesign of refill and flush mechanisms.

In some cases, these changes have been driven by the need to conserve water, a vital resource. In others, the changes have been market and style driven.

Take tank capacities, for example. In bygone years, tanks were mounted on the wall above the bowl. Those old-timers used 5 gal. or more per flush. New designs have brought this down to 1.6 gal., and they're getting smaller. In addition, many new refill and flush mechanisms have also appeared on retailer's shelves. The installation and repair of these devices seem to demand a technical knowledge that doesn't come cheap when you consider labor costs. Armed with some basic skills, however, there are many problems you'll encounter with your toilet that you can fix yourself.

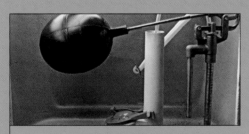

HOW A TOILET WORKS

Flushing and refilling a toilet are accomplished by the interaction of three main components: the flush mechanism, the refill valve, and the float.

Pushing the handle down raises the flapper (actually a seal in the bottom of the tank), and that allows water to escape from the tank and flow into the bowl, where it creates a siphon effect that pulls waste water into the main drain ❶. As the tank water level drops, so does the float (a plastic or metal ball or other device), an action that opens the refill valve and lets water reenter the tank from the supply line ❷. At the lower water level, the flapper reseals the drain in the bottom of the tank and the water from the refill valve fills the tank again ❸. The float now rises and coming to a stop at the end of its upward travel, shuts off the water in the refill valve, making the tank ready for the next flush ❹.

There are many different types of refill valves but they all do the same thing: help flush the bowl and fill the tank with fresh water. Typically this is accomplished in four steps.

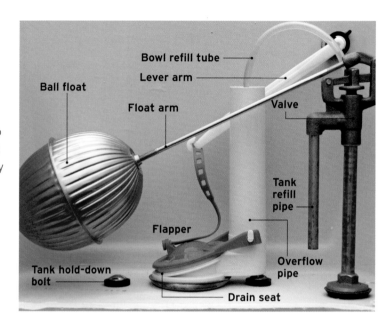

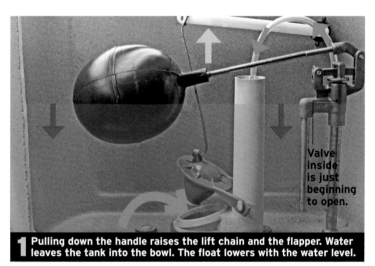

1 Pulling down the handle raises the lift chain and the flapper. Water leaves the tank into the bowl. The float lowers with the water level.

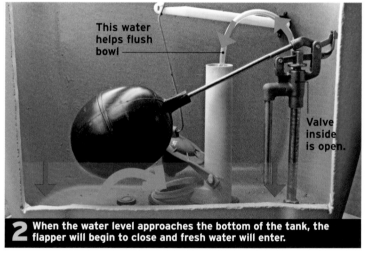

2 When the water level approaches the bottom of the tank, the flapper will begin to close and fresh water will enter.

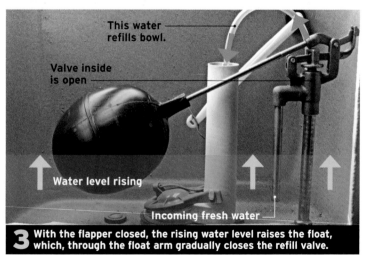

3 With the flapper closed, the rising water level raises the float, which, through the float arm gradually closes the refill valve.

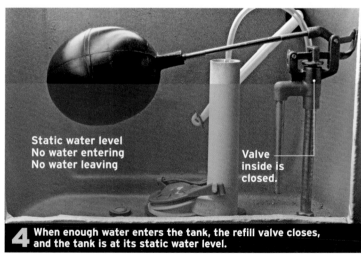

4 When enough water enters the tank, the refill valve closes, and the tank is at its static water level.

DRAINING A TOILET TANK AND BOWL

Many toilet repairs will require you to drain the tank and bowl. You may believe you can get by without this step, (and some repairs don't actually require it). But for repairs that do, taking the time to remove standing water in the tank and bowl will save you from working in and cleaning up an aggravating mess. From performing a simple repair to winterizing a house, the bottom line is the same: to get the water out. However, avoid using metal objects to bail out the water—the metal will leave gray marks on the porcelain.

Turn the tank stop valve off ❶. If the valve is stuck, use adjustable pliers to loosen the nut under the handle. Close the valve and retighten the nut. Check and fix leaks before you go any further with your repair. Flush the toilet and hold the handle down several seconds longer than normal to drain the tank ❷. Then, using an old towel, mop out any left over water in the tank ❸. Wring out the towel and push it back in to the tank again, repeating the process to get as much water out as possible. Be careful not to put pressure on the tank mechanisms or bend them. Then using a small plastic container, dip the water out of the bowl ❹. Something pliable, like an empty margarine or cottage cheese container works well. Dip out as much water in the bowl as you can. No matter how much bailing you do, however, there will be a couple inches of water left in the bowl. Use a rag or an old towel to absorb it ❺.

1 Turn off the water to the tank.

2 Flush the toilet, holding the handle down longer than normal.

3 Using an old towel, absorb any left over water in tank. Get as much water out of the tank as you can.

4 Using a small plastic container, dip the water out of the bowl.

5 Push an old towel into the bowl to absorb any remaining water.

WHAT CAN GO WRONG

If the stop valve won't turn off, loosen the packing nut (the nut under the handle) and turn the valve down. Retighten the nut. If that fails, you'll have to shut off the water at the main house valve and replace the toilet shut-off valve. Do this before going any further with your toilet repair.

INSULATING A SWEATING TANK

A toilet tank sweats when the cold water condenses moisture from the surrounding air. And the water from a sweating tank can rot a floor. Before you fix this problem, make sure the symptoms aren't caused by a leak. They are often indistinguishable.

There are two primary techniques for curing a sweating tank. You either have to raise the temperature of the incoming water or break the air/porcelain contact. Unfortunately most of the solutions, including those shown here, will not prove problem free. You can also install a tempering valve (it mixes a little hot water with the cold before it enters the tank), but these valves have a habit of going bad in a couple years or so. The best solution is a fabric slip-on cover. Sadly, slip-on covers are not sold anywhere—you have to make your own.

To install thin foam insulation from a kit, drain the tank, towel it out, and let it dry.

➡ See "Draining a Toilet Tank and Bowl," p. 113.

Then cut the foam to fit ❶ and glue it to the inside of the tank ❷. A messy alternative that can work if properly applied is automobile undercoating. Start with a tank interior that's clean and dry and wait 24 hours for the coating to cure. If you run the water line to the toilet along the hot water line for 20 ft. (and insulate both), it will raise the water temperature enough so the tank will not sweat.

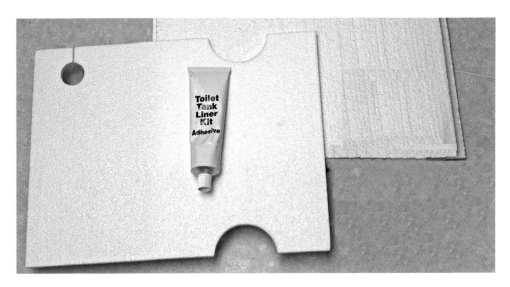

Kits containing thin foam insulation are available for fixing a sweating tank. Make sure the tank is empty, clean, and dry.

1 Measure the tank carefully and use a utility knife to cut the polystyrene to fit.

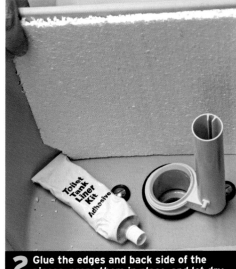

2 Glue the edges and back side of the pieces, press them in place, and let dry.

Automotive spray undercoating can provide a good but messy solution. Don't clog any valves. It may take 48 hours to dry.

RAISING THE TEMPERATURE

Pairing the hot water line along the cold water line to the toilet and insulating both of them can often raise the water temperature in the tank enough to stop sweating.

INSTALLING A BOLT-ON TOILET SEAT

Bolt-on toilet seats have been standard installation for just about as long as toilets have been around. Replacing a bolt-on unit can be accomplished in a matter of minutes. The only aggravation it may cause is removing and/or tightening the nuts on the underside of the bowl flange, but that can be accomplished with the right tools.

If the lid hinge is metal (and for some plastic hinges too), the replacement seat will come with washers to be placed between the hinge and the bowl flange surface ❶. With the washers in place, position the hinge over the holes and insert the bolts through the washers and flange holes ❷. From underneath, twirl on the nuts and tighten them down ❸. You may luck out and be able to tighten the nuts with a pliers ❹. You can also replace the nuts with wing nuts of the same size to make the job easier. If the bolts are in a narrow location making it hard to get your fingers on them, tighten them with a toilet-nut tool. You can pick one up at most hardware stores.

1 Set the washers (if any) over the holes on top of the rear bowl flange.

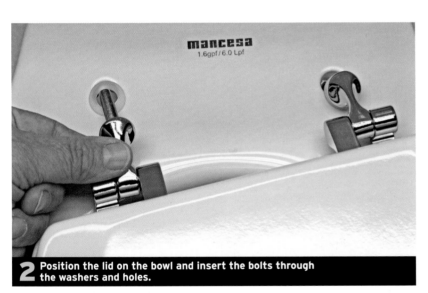

2 Position the lid on the bowl and insert the bolts through the washers and holes.

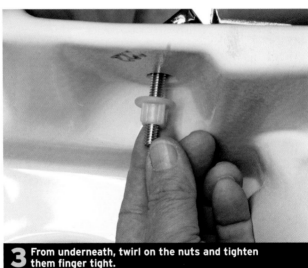

3 From underneath, twirl on the nuts and tighten them finger tight.

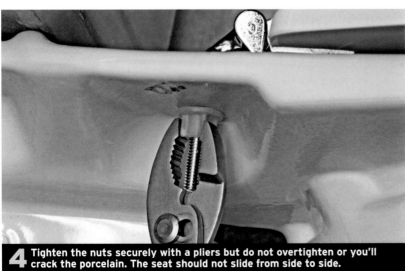

4 Tighten the nuts securely with a pliers but do not overtighten or you'll crack the porcelain. The seat should not slide from side to side.

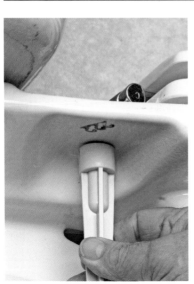

If needed, use a special toilet-nut tool, **which makes tightening the lid nuts easier.**

INSTALLING A SNAP-ON TOILET SEAT

A snap-on toilet seat lifts off its mounts to make cleaning easy. Once you clean the toilet, you just snap it back in the mounts. No tools are required. All this is accomplished with plastic pancake bolts mounted on the toilet bowl. The caps, mounted on the lid, lock the unit in place when inserted in the bolt head and turned.

Before mounting the pancake bolts, grease the inside of the caps with plumber's grease or Vaseline® to make them easier to turn. Spin the bolt head in the caps to spread the grease evenly ❶. Insert the bolts in the seat-mounting holes in the toilet bowl ❷. Working from underneath, tighten the bolts securely with the manufacturer's nuts ❸. If you're fingers won't fit in the recess, tighten the bolts with a pliers or toilet-seat nut wrench.

➡ See "Installing a Bolt-On Toilet Seat," p. 115.

Place the lid on the bowl, pushing the seat caps onto the pancake heads ❹ and pushing both caps to the right to lock them ❺.

Spread grease in the cap by twirling a bolt in it.

1 Squeeze plumber's grease or Vaseline into the seat caps. Twist a bolt head in each cap to spread the grease thoroughly.

2 Insert the pancake-head bolt into the holes in the bowl.

3 From underneath, use the supplied nuts and tighten them down.

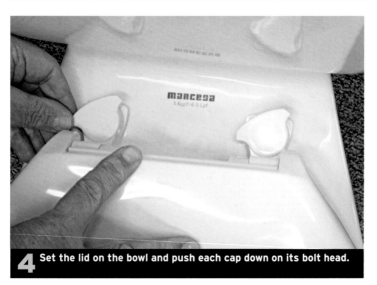
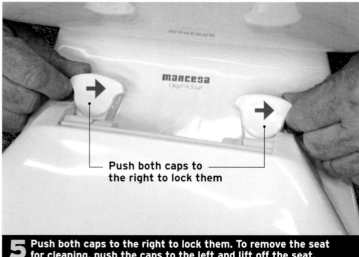

Push both caps to the right to lock them

4 Set the lid on the bowl and push each cap down on its bolt head.

5 Push both caps to the right to lock them. To remove the seat for cleaning, push the caps to the left and lift off the seat.

REPLACING A TOILET TANK HANDLE

A broken plastic toilet handle is easy to replace. However, when reconnecting the plastic or metal flapper chain to the handle, you may have to cut or bend the extension arm inside the tank. Tank handles are typically chrome, white, brushed nickel, or polished brass. Do not get a polished brass lever unless the finish is guaranteed for life.

First, disconnect the valve chain from the extension arm ❶. Then loosen and remove the holding nut by turning it clockwise; it is a left-hand thread ❷. Remove the handle by pulling it straight out the front ❸. Insert a new handle and turn the nut counterclockwise. Reconnect the handle to the flush valve chain at exactly the same location it was removed. You may have to bend or cut the extension arm to correct its position.

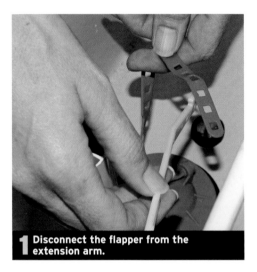

1 Disconnect the flapper from the extension arm.

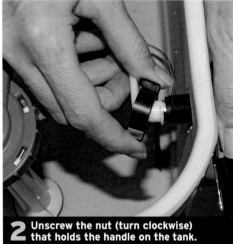

2 Unscrew the nut (turn clockwise) that holds the handle on the tank.

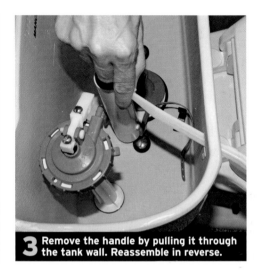

3 Remove the handle by pulling it through the tank wall. Reassemble in reverse.

REPLACING A BALL FLOAT

B all floats occasionally fill with water, won't rise when the tank refills, and cannot turn the fill valve on and off properly. To test, mark the existing float's location on the rod with tape ❶, unscrew the ball ❷, and shake it. If you hear water, that's bad. If the rod starts to turn as well, use a pliers to hold it in place. Once you've removed the old float, simply screw on a new one up to the marked spot ❸.

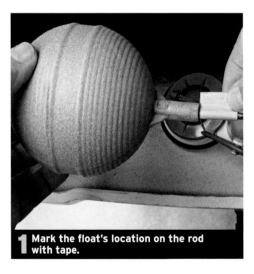

1 Mark the float's location on the rod with tape.

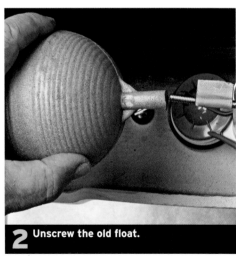

2 Unscrew the old float.

3 Screw a new float back on the rod, stopping at the tape. Remove the tape.

ADJUSTING REFILL MECHANISMS

Any refill mechanism has to do two things. First, It has to sense the lack of water in the tank and allow water to enter. After that, it has to sense the amount of water in the tank and close the valve at the right water level. Rarely do mechanisms come properly adjusted right out of the box. Once you install a toilet or change out the refill mechanism, you have to run it through several cycles to verify whether it's adjusted properly. Not enough water in tank will result in a poor flush. Too much causes the toilet to run constantly. You want the valve to shut off when the water is around $\frac{1}{2}$ in. below the top of the overflow tube. Depending on the model, all this is done by moving something up and down or by turning a screw or knob.

Adjusting a ball-and-float valve

If the water level is too low, and there's no adjustment screws on your refill mechanism, grab the float arm with both hands and place your thumbs on top of the arm about 3 in. forward of where it screws into the valve. Bend the float arm up slightly ⒜. This will raise the float ball and cause it to let more water in the tank before shutting it off.

A water level higher than $\frac{1}{2}$ in. from the top of the overflow tube is okay, as long as the water stops before it reaches the top of the tube. If the water rises higher than that, you will have to bend the float arm down slightly to reduce the amount of water in the tank ⒝.

Certain valve designs incorporate float-arm adjustment screws. Adjust the screw to raise or lower the float arm to control the water level ⒞.

Adjusting a Fluidmaster® valve

Fluidmaster valves use a knurled and slotted knob to adjust the water level. For any adjustment, turn the knob one small rotation, then flush and readjust if necessary. If the water level is too low, turn the knob clockwise (with your fingers or a screwdriver) to allow more water into tank. ⒟ If the water is too high, turn the knob counterclockwise. ⒠ You may have to repeat this procedure several times before you get it adjusted right.

With any type of refill valve, you want the float to shut the water off about $\frac{1}{2}$ in. below the top of the overflow tube.

Correct water level: about $\frac{1}{2}$ in. below top of overflow

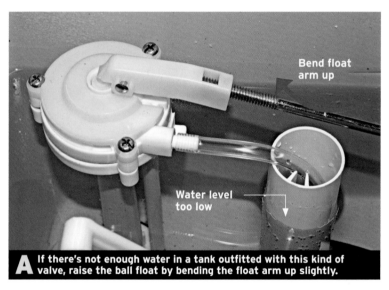

Bend float arm up

Water level too low

A If there's not enough water in a tank outfitted with this kind of valve, raise the ball float by bending the float arm up slightly.

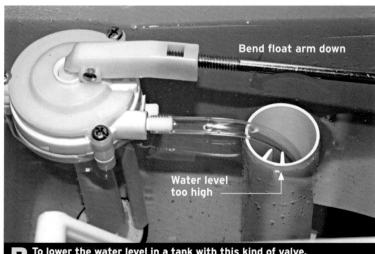

Bend float arm down

Water level too high

B To lower the water level in a tank with this kind of valve, bend the float arm down slightly.

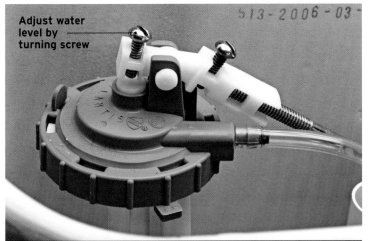

Adjust water level by turning screw

513-2006-03

C If an adjustment screw is present on your valve, flush the toilet several times and adjust the front screw to get the level right.

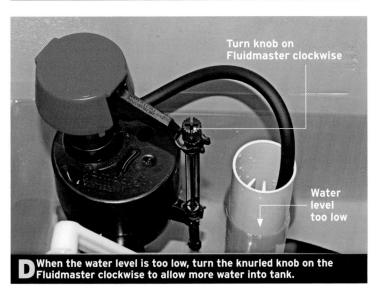

Turn knob on Fluidmaster clockwise

Water level too low

D When the water level is too low, turn the knurled knob on the Fluidmaster clockwise to allow more water into tank.

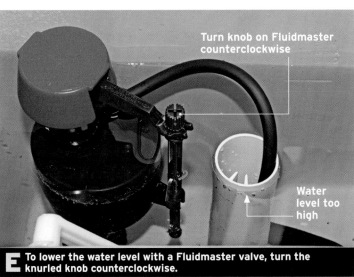

Turn knob on Fluidmaster counterclockwise

Water level too high

E To lower the water level with a Fluidmaster valve, turn the knurled knob counterclockwise.

Vertical column refill tube

This mechanism has no outside adjustments. There is a float inside the housing on the mechanism. On the housing is a line marking the ideal water level. Adjust the water level to that line by twisting the upper end, adjusting its height, and then twisting in the opposite direction to lock it in place.

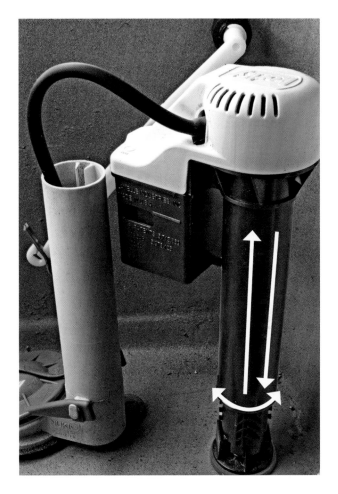

REPAIRING A BOLT-DOWN REFILL VALVE

1 Turn the water off and drain the tank. Remove the ball float arm and ball float, if necessary, by unscrewing the arm from the valve head.

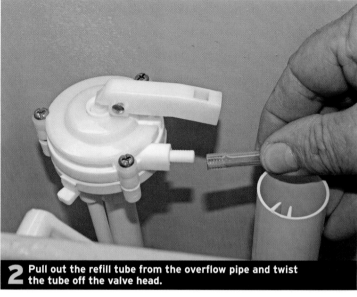

2 Pull out the refill tube from the overflow pipe and twist the tube off the valve head.

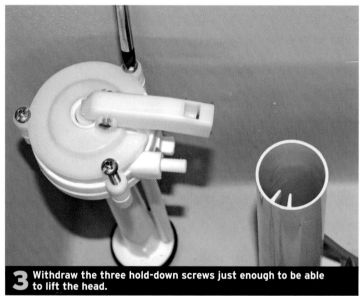

3 Withdraw the three hold-down screws just enough to be able to lift the head.

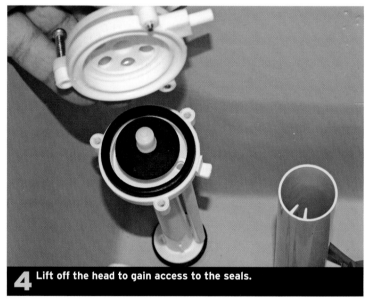

4 Lift off the head to gain access to the seals.

A bolt-down fill valve is actuated by the up-and-down travel of a ball float. As the ball raises and lowers, it closes and opens a rubber seal within the bolt-down cap. The cap houses a seal and an O-ring, and if you're getting leaks at the juncture of the cap and body, one or both of these needs attention.

Repairing any bolt-down system starts by removing the float ball and arm. In most cases, this will come away as a unit by un-screwing the arm (with pliers if necessary) **❶**. If there is a clearance problem and you cannot remove the ball and arm together, remove them separately. Don't bend the arm to give you enough clearance to remove it. You'll change the adjustment and will have to readjust after reassembling it. (You may anyway, of course.) Then pull the refill tube up out of the overflow pipe and off the valve head **❷**. Using a Phillips screwdriver,

unscrew the three screws that hold the head on the vertical column **❸** and lift the head off **❹**. Do not lose the screws. Remove the center seal and look for cuts on it, as well as sediment that might be clinging to it **❺**. Clean the seal or replace it. The replaceable O-ring keeps the water inside the head of the refill valve. If water was leaking from the seam between the cap and body, replace the O-ring **❻**.

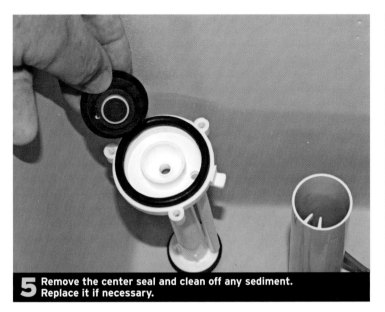

5 Remove the center seal and clean off any sediment. Replace it if necessary.

6 Remove the O-ring and replace it if necessary.

REPAIRING A BRASS REFILL VALVE

You do not have to remove the ball float to work on a brass refill valve. You will see flat tabbed screws holding the arm to the refill head. Grab the flat of each screw with a needle-nose pliers and remove the screws **❶**.

> **See "Don't Lose the Parts," p. 97.**

Lift out the assembly with the float arm attached **❷**. Pull the valve stem straight up and out of the valve body **❸**. Check the bottom of seal for cuts and clean off any sediment. Replace the O-rings mounted around the side of the valve stem, as well as the flat seal on the bottom.

Rear pivot screw

Front pivot screw

1 Turn the water off and drain the tank. Using needle-nose pliers, unscrew the pivot screws, front and back.

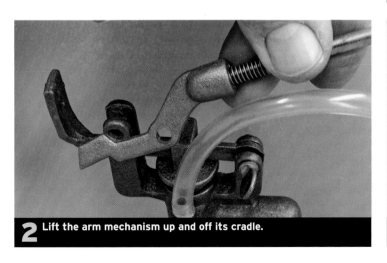

2 Lift the arm mechanism up and off its cradle.

3 Pull the valve stem up and out. Clean or replace the O-rings and bottom seal.

REPLACING A GENERIC REFILL VALVE

Regardless of refill valve installed in your toilet, eventually you'll have to replace it. Nothing lasts forever, certainly nothing that sees as much action as a refill valve. The good news is that all refill valves come out basically the same way. You start, of course, by turning off the water at the stop valve and disconnecting the supply tube from the refill valve. Then you drain the tank, and you're ready to go.

➡ **See "Draining a Toilet Tank and Bowl," p. 113.**

Working from the underside of the tank, you'll see a large plastic nut on the threads of the refill valve where the body comes through the bottom tank wall. Remove the nut with a pliers or tongue-and-grooved pliers **❶**. All refill valves, regardless of type, will pull straight up and out of the tank once you've removed the holding nut. So will yours. Lift out the valve and make sure you find the seal at the bottom **❷**. Discard everything including the seal. The new refill valve must seal at the bottom where the threads go through the tank. Some valves come with the seal attached to the body. With others, you'll have to slip the seal (found in the valve replacement kit) on the threads yourself **❸**. With the seal on the bottom of the valve body, insert the threaded male end into the hole in the tank **❹**. Reaching over the top of the tank, hold the refill valve in place and tighten the holding nut finger tight. Continue holding the valve while you make the last few tightening turns with pliers, or else the valve body will spin **❺**. If your replacement refill requires a ball float mechanism, screw it on now **❻**. Next comes the refill tube. The refill tube may be separate and you will have to push it onto the mini threads of the new refill valve. On most models, these tubes will be too long and have to be cut to fit. Push the refill tube onto the orifice in the valve (you may have to twist is back and forth a bit while you're pushing) and cut it to length. Insert the cut end into the overflow **❼**. Finally, attach the supply tube and tighten with pliers **❽**.

➡ **See "Adjusting Refill Mechanisms," p. 118.**

1 On the bottom of the tank, unscrew the large nut which holds the fill valve body in place.

4 Insert the replacement valve in the hole in the bottom of the tank.

7 Insert the refill tube on valve and cut it to length. Once cut, insert the refill tube into the overflow pipe.

2 Pull the refill valve up and out of tank.

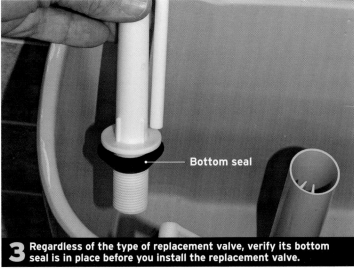

Bottom seal

3 Regardless of the type of replacement valve, verify its bottom seal is in place before you install the replacement valve.

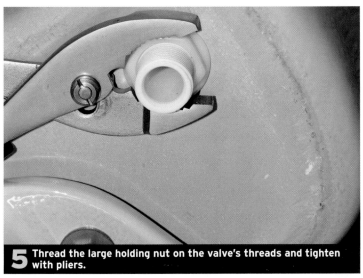

5 Thread the large holding nut on the valve's threads and tighten with pliers.

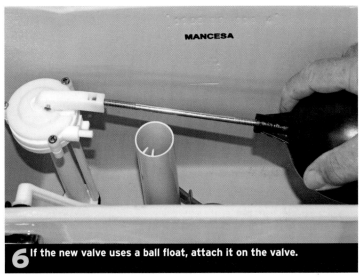

6 If the new valve uses a ball float, attach it on the valve.

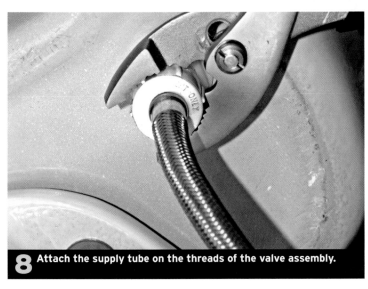

8 Attach the supply tube on the threads of the valve assembly.

TRADE SECRET

The interior of your toilet tank will show you the condition of your water supply. Fine black dust means you are pumping in mineral sediment from the well. Red? You have iron in your water. Red and slimy? This is iron bacteria. Green? Your copper is being eaten away by chemicals in the water. You can have your water tested by a service specializing in these techniques and get recommendations on corrective action you can take, if necessary.

REPAIRING A TWIST-OFF HEAD

The cap on a lock-down refill valve is not held in place with screws, but with tabs under the head. Like the bolted model, you'll find it used on ball-and-float systems. The cap twists off the body after you remove a locking tab.

First, turn the water off and drain the tank, then unscrew the float arm and remove it. Pliers will give you the leverage you need if the arm proves difficult to unscrew. Then pull the refill tube out of the overflow pipe ❶. Look for a vertical tab on the top outside rim of the cap. That's a locking key. Grab the tab with needle-nose pliers and pull straight up ❷. Set the tab aside and don't lose it. Twist the top of the refill mechanism until you can see white in the slots of the cap ❸,❹. Then pull the cap off the body ❺. With the head off, you can now check for sediment on the seal and inside the white vertical tube. Also check the seal to see if it has been cut and that its mating surface on the white tube is also smooth and uncut. Replace the seal if it's damaged. If the refill mechanism was leaking water just under the flat head, replace the O-ring seal ❻.

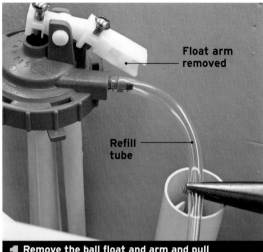

Float arm removed

Refill tube

1 Remove the ball float and arm and pull the refill tube out of the overflow pipe.

2 Using pliers, grab the vertical locking tab and pull it straight up.

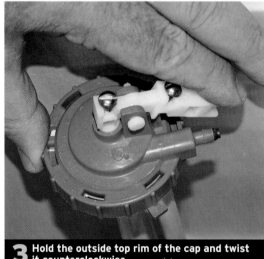

3 Hold the outside top rim of the cap and twist it counterclockwise.

4 Untwist the cap until the white tabs completely fill the blue slots in the cap.

5 Pull the cap off and examine the inside center seal. Replace it if it's damaged.

6 Remove the O-ring from the column and replace it to stop leaks from the cap.

INSTALLING/ADJUSTING A FILLPRO REFILL VALVE

The Fillpro refill valve is the smallest and most unusual of all the refill valves. It uses an air-pressure differential to determine when to turn the water off and on. Because it senses the air pressure external to the tank by means of a slot in the holding nut under the tank, you have to use the manufacture's nut to attach the valve. One other caveat comes with Fillpro installation—don't tighten the nut so much that it crushes the spacers between the air slots. The Fillpro is not a maintainable valve. If it stops working, replace it.

The Fillpro installs like every other fill valve. First, drain the toilet and the tank.

➤ See "Draining a Toilet Tank and Bowl," p. 113.

Then remove the old valve, insert the Fillpro threads through the hole in the bottom of the tank, and tighten the nut down from underneath ❶. Insert the refill tube into the overflow pipe, connect the water supply, and turn the water on ❷. Flush the toilet several times, adjusting the level with the knurled plastic knob on top of the valve ❸. Turn the knob clockwise to raise the water level and counterclockwise to lower it.

➤ See "Replacing a Generic Fill Valve," p. 122.

➤ See "Adjusting Refill Mechanisms," p. 118.

1 Shut off the water, drain the tank, and remove the old valve. Install the Fillpro valve.

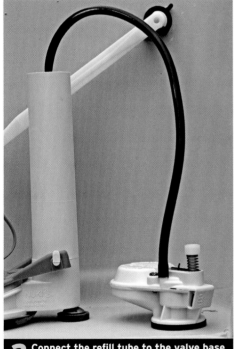

2 Connect the refill tube to the valve base and insert it in the overflow tube.

Slot

This slot in the hold-down nut on the tank bottom allows the unit to determine the air-pressure differential. It must not be obstructed.

3 Flush the toilet several times and adjust the water level with the spring-loaded knob on the base of the unit.

REPLACING A FLUSH VALVE

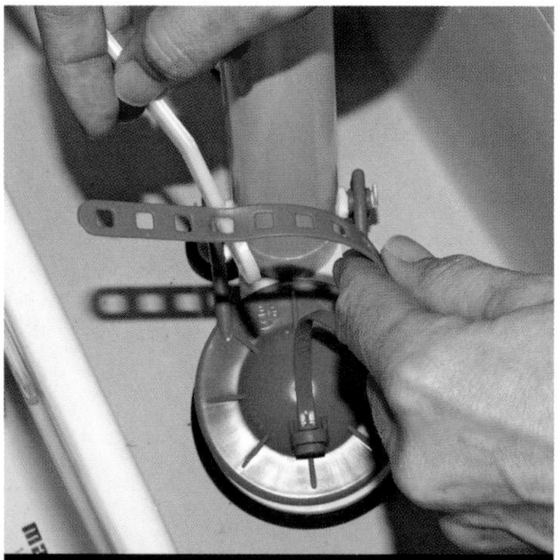

1 Disconnect the flapper or flush valve from the flush handle end.

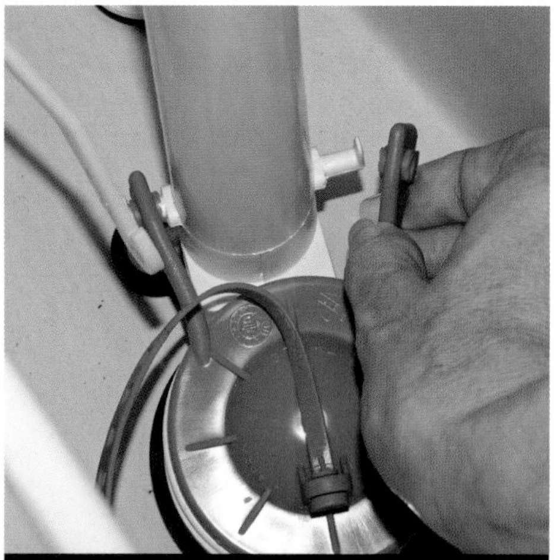

2 Pull the tabs of the flapper valve off the posts on the overflow pipe and discard.

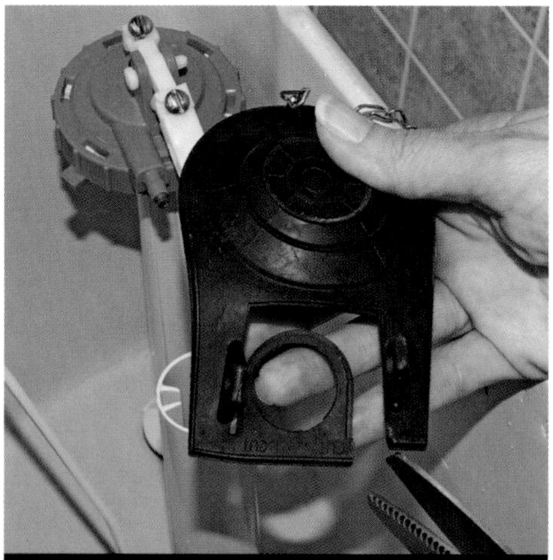

3 Cut out the circular center from the replacement flapper with a scissors.

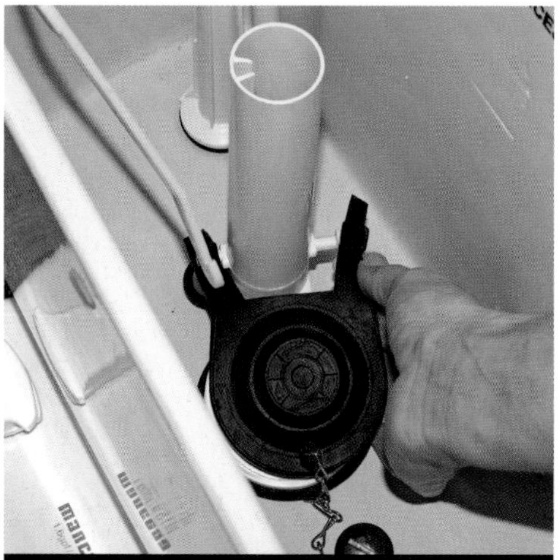

4 Slip the tabs of the new flapper on the posts of the overflow tube.

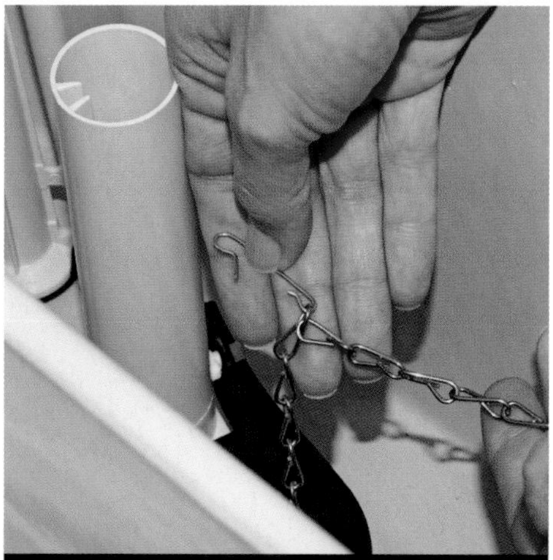

5 Insert the hook on the chain about halfway up the chain.

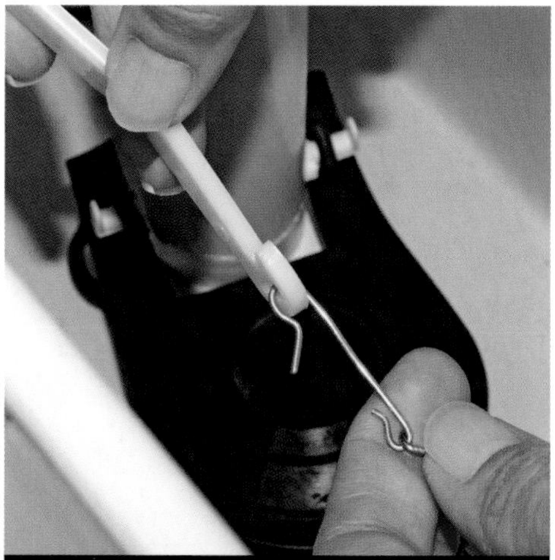

6 Connect the opposite end of the hook to the handle arm. Adjust as necessary.

Replacing a flush valve (flapper) is perhaps the most common repair procedure performed on an old toilet. Over time and due to the effects of chlorine and minerals in the water, the flexible flapper that seals the drain in the center of the tank becomes rigid. When this happens, water will start seeping out of the tank and you will hear it occasionally refilling on its own. Fortunately, replacing a flapper is an easy task. Make sure, however, to purchase a high-quality flapper that boasts of staying flexible in hard water. Then, as with all toilet repairs, turn the water off, and flush the toilet to be rid of the water in the tank. Mop up the remaining water with an old towel. Have some paper towels handy; sometimes the black coloring on the flapper will soil your fingers.

The first step in replacing a flapper is removing the old one. First, detach it from the tank lever ❶. Note where it is connected to the lever or tape the location. The new one should connect to the same spot. Pull off the tabs that connect the flapper from their posts on the bottom of the overflow tube. Discard the flapper ❷. If your replacement comes with a circular center, cut it out with a sharp scissors, leaving only the tabs ❸. Attach the tabs of the new flapper to the posts on the bottom of the overflow tube ❹. Insert the hook about halfway up the chain ❺, and connect the opposite end of the hook to the handle arm. Push the handle down to see if the handle raises the flapper fully at the same time the handle would hit the tank lid, or just before. Adjust the position of the hook on the chain until it does ❻. As the handle is released, the flapper must close fully but still leave some slack in the chain. Plastic chains will connect directly to the handle but the logic of opening and closing remains the same. When you think you have the chain adjusted properly, turn the water on and let the toilet refill. Adjust the chain so the flapper opens fully and seats firmly in the drain hole when closing.

INSTALLING A FLAPPER-VALVE SEAT

This Fluidmaster 555C repair kit can save you the trouble of taking the tank off and replacing the entire mechanism.

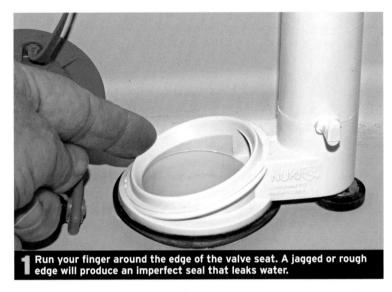

1 Run your finger around the edge of the valve seat. A jagged or rough edge will produce an imperfect seal that leaks water.

I f the flapper valve doesn't seat properly, the toilet will constantly try to refill itself, running all the time and wasting water. Although sometimes a hardened flapper valve causes the problem, the culprit is often that the surface of the bottom drain seat is roughened and eroded from sediment or chemicals. To determine where the problem lies, turn the water off and flush the toilet to drain the tank. Run your finger around the edge of the flapper seat ❶. It should feel completely smooth. If the edge feels rough, you normally have to replace the entire overflow mechanism which is a lot of messy work.

➡ **See "Replacing the Tank Gasket and Overflow Pipe," p. 129.**

As an alternative, you can install a Fluidmaster 555C kit (it takes about 3 minutes) which comes with a form-fitting putty to seal the valve seat smoothly, as well as a new flapper.

Start by turning off the water to the toilet and flushing out the tank. If your system employs a float ball and rod, remove them ❷. Tape the position of the float ball on the rod and unscrew the float ball.

➡ **See "Draining a Toilet Tank and Bowl," p. 113.**

➡ **See "Replacing a Ball Float," p. 117.**

Then use a pliers to unscrew the float arm from the valve. Set both of these parts aside. Unhook the chain from the lever arm and discard it as well as the old flapper ❸. Remove the putty from the repair package, keeping it perfectly round. Center the putty on the valve seat and

>> >> >>

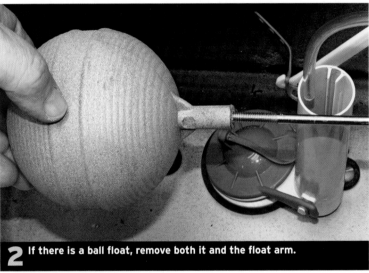

2 If there is a ball float, remove both it and the float arm.

3 Remove the flapper or flush valve.

INSTALLING A FLAPPER-VALVE SEAT (CONTINUED)

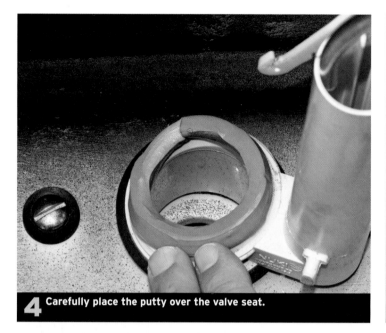

4 Carefully place the putty over the valve seat.

5 The bottom of the new valve will form a concave seal in the putty.

6 Install the valve directly under the chain from the lever arm.

7 Connect the chain to the lever arm.

push it in place **4**. The bottom of the new valve is slightly concave and will make a conforming seal with the putty **5**. Before pressing the new flush valve in place, note the proper angle of the flush valve. Typically, it has to be to the side of the of overflow pipe below the chain from the lever arm **6**. Review the manufacturer's instructions. Connect the chain to the lever arm **7**. Replace the float ball and arm. Turn the water back on and make sure the float valve opens and closes properly by flushing the toilet several times. Adjust the chain and ball float if necessary.

➤ See "Adjusting Refill Mechanisms," p. 118.

REPLACING THE TANK GASKET AND OVERFLOW PIPE

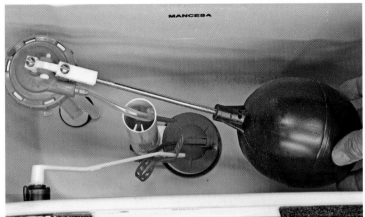

1 If your tank uses a ball float, removing the float and arm isn't absolutely necessary, but it will make the job go faster.

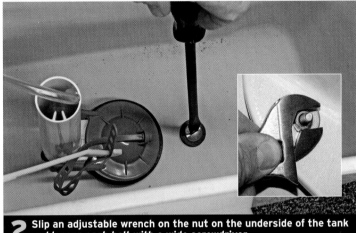

2 Slip an adjustable wrench on the nut on the underside of the tank and loosen each bolt with a wide screwdriver.

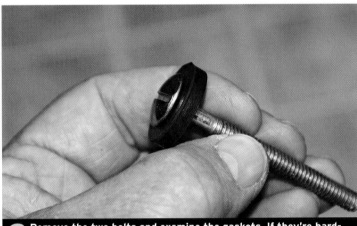

3 Remove the two bolts and examine the gaskets. If they're hardened or deteriorating, replace them when reassembling the tank.

4 Grabbing the tank on each side, pull it straight up and off the base.

Every now and then an overflow pipe will crack or the flapper valve seat will become so ragged it needs to be replaced. Both conditions will create a leak under the flush valve or between tank and bowl and will require replacement of the overflow pipe and tank gasket.

To start the repair, drain and dry out the tank.

> **See "Draining a Toilet Tank and Bowl," p. 113.**

If your toilet has a float and arm, tape the position of the float on the arm, remove both, and set them aside. The tape will tell you how far to screw the float back on when reassembling **1**.

> **See "Replacing a Ball Float," p. 117.**

Remove each of the two bolts holding the tank to the base and examine the condition of the gasket under the bolt **2**. If either gasket has hardened or is breaking apart, you'll need to replace both bolts and gaskets **3**. They come as a set at all hardware stores. With the bolts removed, the tank is loose, and you can remove it by grabbing it on both sides and pulling straight up **4**. Set the tank

>> >> >>

REPLACING THE TANK GASKET AND OVERFLOW PIPE (CONTINUED)

5 Gently set the tank on the floor or rug and pull off the tank gasket.

6 Loosen the nut holding the overflow pipe and then turn it by hand and remove it.

7 Remove the refill tube from the overflow pipe.

8 Lift the overflow assembly up and pull it straight out. Set the replacement overflow pipe in the hole in the bottom of the tank.

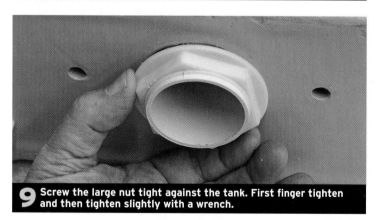

9 Screw the large nut tight against the tank. First finger tighten and then tighten slightly with a wrench.

10 Push the foam seal over the threads and up against the tank bottom.

11 Slip the tank bolts with gaskets into the holes and tighten them from underneath.

on a rug on the floor, and remove the tank gasket. If the tank was leaking around the seal, replace the gasket **5**. Then, using a large pipe wrench, loosen the nut that holds the overflow pipe to the tank **6**. Remove the nut and set it aside. If the wrench won't turn the nut, split it with a sharp chisel to remove it. Turn the tank top-side up and remove the refill tube from the overflow pipe by freeing it from its clip or unsnapping it from its grooved housing **7**. Pull out the overflow

assembly and replace it with the model of your choice **8**. You'll find replacement assemblies at all hardware stores. Turn the nut over the threads and against the tank bottom and tighten it with a wrench **9**. Push the seal on the overflow pipe and set the tank on the bowl with the seal in the hole **10**. Insert new tank bolts with gaskets and tighten the nuts under the tank flange **11**. You can replace these nuts with wing nuts to make this job easier, but do not tighten them exces-

sively. Level the tank if necessary. Place the refill tube in the overflow tube and replace the float and arm. Connect the supply line to the tank, turn the water on and adjust the water level. Check for leaks.

➤ See "Adjusting Refill Mechanisms,"
p. 118.

TROUBLESHOOTING A TOILET

Symptom	Cause	Solution
Tank constantly refills	1. Water is leaking out the bottom seal into the bowl. This can be caused by a bad flapper (hardened over time) or flush mechanism.	Replace the flush mechanism or flapper.
	2. The drain seat is rough and deteriorated, preventing the flapper from seating properly.	Replace the drain seat or install a Fluidmaster 555C.
	3. Cracked overflow pipe. The vertical pipe has deteriorated, preventing the flapper from seating properly.	Replace the overflow pipe or complete overflow assembly.
	4. The refill valve or float is misadjusted.	Adjust the refill valve so the water turns off about 1/2 in. below the top of the overflow pipe. If not possible, replace the refill mechanism or valve.
	5. The flapper chain is too tight causing the flapper to be above it's seal.	Add slack to the chain.
Nothing happens when handle is pushed down	1. The connection from the handle to the chain has come loose or water is not turned on.	Reconnect, replace, or adjust the chain. Make sure water is turned on.
	2. There is too much slack in chain.	Tighten the chain.

REPLACING A TOILET

When you've grown tired of the style or the faulty operation of your old toilet, or decide to remodel your bath, it's time to replace the toilet. The techniques shown here can also be used if you need to remove the existing toilet to repair the floor under it.

To keep from getting yourself and the floor wet, turn the water off to the tank. Then drain both tank and bowl completely. This also makes the unit lighter to carry.

➡ See "Draining a Toilet Tank and Bowl," p. 113.

Using a small pipe wrench or pliers, loosen the large nut on the bottom of the tank that connects the supply tube to the refill mechanism ❶. To disconnect the bowl from the floor, you must remove the nuts and washers holding the tank to the toilet flange ❷. Place an old rug on the floor in front of the toilet. Then, wearing gloves, straddle the toilet, grab each side of the tank, and keeping your back straight, lift straight up with your knees to break the bowl free from the floor ❸. Set the toilet down on the old rug, grab the rug, and pull the toilet out of the way ❹. Do not do this if you have a bad back. If the toilet proves too heavy, disassemble the tank from the base and remove the two pieces separately.

➡ See "Replacing the Tank Gasket and Overflow Pipe," p. 129.

To keep sewer gas from entering the house and parts or debris from entering the drain, plug the drain hole with a crumpled rag ❺. Then remove the old wax seal from the flange. Grab it with pliers, yank it up from the flange, and discard it ❻. In most cases, you will knock the bowl hold-down bolts to the side as you pull the old toilet off the floor. If this is the case, straighten them at this time. As shown here, you can anchor the bolts straight with a nut securing them to the flange ❼. Because wax from the seal seems to get on everything if it is not cleaned up completely, do a second cleaning if necessary. Get as much up as you can, using wood shims and throwaway rags ❽. Insert a new wax ring on the toilet flange or horn ❾. With a level, verify that the floor is flat. If not, use shims around the edges where the toilet will sit.

➡ See "Installing a Bolt-On Toilet Seat," p. 115.

➡ See "Installing a Snap-On Toilet Seat," p. 116.

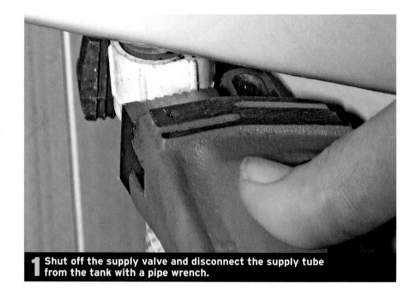

1 Shut off the supply valve and disconnect the supply tube from the tank with a pipe wrench.

4 Set the toilet down on the rug. Pull the rug and toilet along the floor and outside.

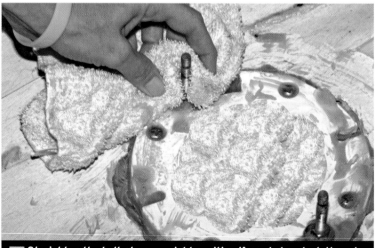

7 Straighten the bolts to an upright position if needed, or lock them to the flange with nuts.

2 Remove the cap and nuts from the bowl hold-down bolts.

3 Straddle and lift the toilet straight up off the flange bolts. Have an old rug handy.

5 Plug the drain hole with a large rag and scrape up wax off floor.

6 Using pliers, pull the old seal from flange and discard it.

8 Do a second cleaning around flange if needed.

9 Insert the new wax ring on the flange.

REPLACING A TOILET (CONTINUED)

The next part is tricky—don't do it if you have a bad back. Get someone to help. Lift the bowl above the flange bolts, position the holes over the bolts, and set the bowl straight down on the bolts **10**. Once you have the toilet in place, set a level on the bowl. The bowl does not have to be perfectly level, but a level bowl operates better. Then insert a washer and nut over the hold-down bolts. Tighten the nuts until the bowl is flat against the floor or against the shims **11**. You know it's tight enough when you cannot rock the bowl side to side. Do not overtighten or the bowl may break. Use a screwdriver to push metal shims under the bowl edge or a chisel to snap off any wood shims sticking out from under the bowl. Assemble the overflow tube (if not already installed) and seal and attach the tank to the bowl with the gasketed bolts.

> ➡ **See "Replacing the Tank Gasket and Overflow Pipe," p. 129.**

The tank does not have to be perfectly level, but it should be close **12**. Adjust this level by tightening the tank down slightly more on one side than another. The tank has to be tight enough not to rock. Too tight, and the tank will snap as it grinds against the bowl. Attach the supply tube to the refill threads of the new tank and turn the water back on **13**. Then check for leaks. Be sure to check for leaks on the stop valve itself because just turning the handle or moving the supply tube can cause it to leak. Flush the toilet several times and adjust the water level until it stops about 1/2 in. from the top of the overflow pipe **14**.

> ➡ **See "Adjusting Refill Mechanisms," p. 118.**

Check for leaks around the base of the toilet and tighten the flange bolts slightly if necessary. Orient the lid with its front edge forward and set the tank lid on the tank **15**. Cut the flange bolts off at the proper level so the cap can snap onto the bolt or nut **16**,**17**. Next, you have the option of caulking around the base of the bowl. Use white or clear silicone **18**.

TRADE SECRET

Old toilets have a habit of collecting minerals on the porcelain that slow or stop the water flow. They cannot be fixed. It's time for a replacement.

10 Set the bowl straight down on the flange bolts.

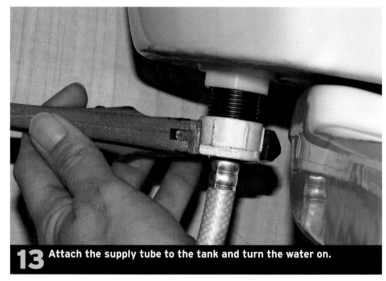

13 Attach the supply tube to the tank and turn the water on.

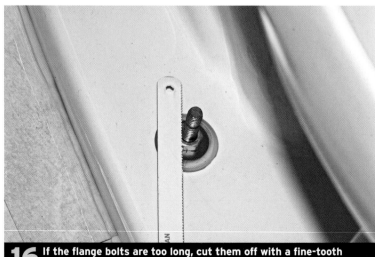

16 If the flange bolts are too long, cut them off with a fine-tooth hacksaw.

11 Gently tighten down the nuts on the flange bolts with an adjustable wrench.

12 Check the level of the tank.

14 Flush the toilet several times, adjusting the water level in the tank. It should come up to within 1/2 in. of the overflow pipe.

15 Gently replace the lid on tank.

17 Fill flange-bolt caps with plumber's putty and install.

18 Seal the base of bowl with clear or white silicone.

INSTALLING A BATHROOM

THE BATH IS THE FIRST ROOM you run to in the morning when you get up and the last you see when you go to bed. Thus having a bath that is bright and cheery can put you in the right frame of mind to start the day.

It is not at all hard to change out a vanity or to install a tub/shower. One thing to keep in mind is that if you're remodeling an existing bathroom, a single-piece shower or tub-shower unit will be too big to maneuver into your existing bathroom space. Instead, buy a shower or tub-shower that consists of several modules designed to interlock. And it is always nice to have the fixtures on hand while you are framing the walls to verify the dimensions they require. More than one shower or tub framing has had to be ripped out due to a wrong dimension.

LAYING OUT A BATHROOM

It is important that the bath be laid out so that there is enough room between fixtures to make them usable. This is especially important if the bath is designed for the handicapped. You must also keep in mind which way the door will swing and mark it on the subfloor as a reminder not to put any fixtures in that area. It is also advisable to put a window in the bath to allow light in and to air it out in the summer.

Where you put the fixtures will depend somewhat on the location of the plumbing in an existing bathroom. Most baths will have a "wet" wall, containing the supply, drain, and vent lines. You'll save yourself a good deal of time and money by locating your new fixtures as close as possible to that wall, avoiding long runs of new pipes.

You can save on hot water use and costs by installing a recirculator, which delivers hot water instantly, even at an end-of-the-run sink.

➡ See "Installing a Recirculator" p. 209.

Installing an accessible bathroom

The needs of persons with disabilities require standards of accessibility other bathrooms would not necessarily incorporate. And although the Americans with Disabilities Act applies primarily to commercial and public facilities, it can provide valuable information for anyone wanting to install an accessible bathroom—new or remodeled. Your local building department should also have guidelines for local codes that apply to residential remodeling. It's important, however, to tailor your installation to the needs of those residing in your home. Don't just blindly follow the rules, especially if strict adherence to them will make your installation more uncomfortable than a custom layout.

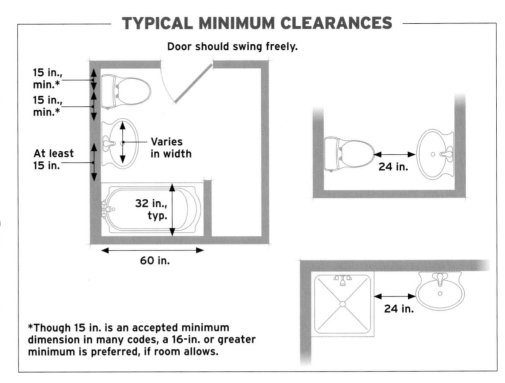

TYPICAL MINIMUM CLEARANCES

Door should swing freely.

15 in., min.*

15 in., min.*

At least 15 in.

Varies in width

32 in., typ.

60 in.

24 in.

24 in.

*Though 15 in. is an accepted minimum dimension in many codes, a 16-in. or greater minimum is preferred, if room allows.

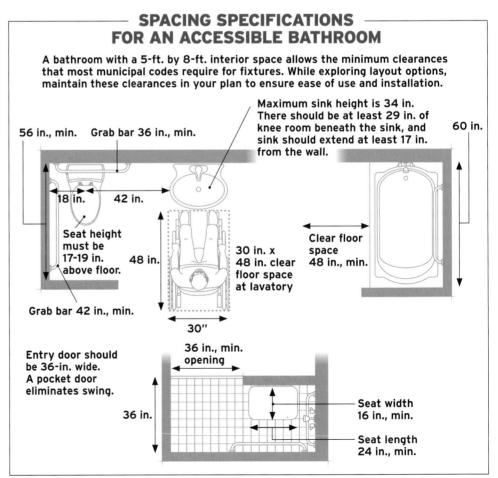

SPACING SPECIFICATIONS FOR AN ACCESSIBLE BATHROOM

A bathroom with a 5-ft. by 8-ft. interior space allows the minimum clearances that most municipal codes require for fixtures. While exploring layout options, maintain these clearances in your plan to ensure ease of use and installation.

Maximum sink height is 34 in. There should be at least 29 in. of knee room beneath the sink, and sink should extend at least 17 in. from the wall.

56 in., min.   Grab bar 36 in., min.

60 in.

18 in.   42 in.

Seat height must be 17-19 in. above floor.

48 in.

30 in. x 48 in. clear floor space at lavatory

Clear floor space 48 in., min.

Grab bar 42 in., min.

30"

Entry door should be 36-in. wide. A pocket door eliminates swing.

36 in., min. opening

36 in.

Seat width 16 in., min.

Seat length 24 in., min.

INSTALLING A PEDESTAL SINK

PEDESTAL SINK ROUGH-IN

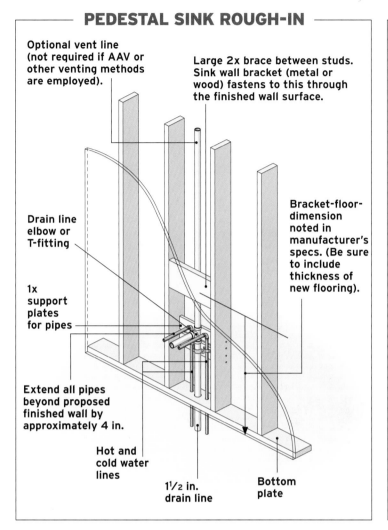

Optional vent line (not required if AAV or other venting methods are employed).

Large 2x brace between studs. Sink wall bracket (metal or wood) fastens to this through the finished wall surface.

Drain line elbow or T-fitting

Bracket-floor-dimension noted in manufacturer's specs. (Be sure to include thickness of new flooring).

1x support plates for pipes

Extend all pipes beyond proposed finished wall by approximately 4 in.

Hot and cold water lines

1½ in. drain line

Bottom plate

SUPPORTING A HEAVY SINK

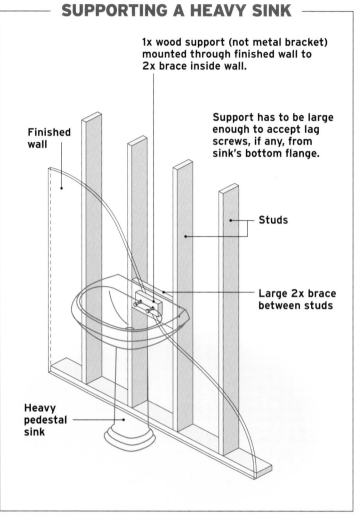

1x wood support (not metal bracket) mounted through finished wall to 2x brace inside wall.

Support has to be large enough to accept lag screws, if any, from sink's bottom flange.

Finished wall

Studs

Large 2x brace between studs

Heavy pedestal sink

Pedestal sinks can bring an element of high style to your bath design, but they come with drawbacks. First, you won't have storage under the sink, and depending on the pedestal design, some or all of the plumbing will be visible. A pedestal sink also requires support in the wall (see illustrations above).

To begin, mark the center of the pedestal on the bottom plate of the wall framing ❶. Then, using the manufacturer's specifications, mark the location of the wall brace and fasten the 2x brace between the studs ❷. Dry-fit the tailpiece, P-trap, and drain extension on the sink, and mark the point where the trap will enter the wall. Fasten a 1x mounting plate at this location

>> >> >>

1 Mark the center of the drainpipe on the bottom wall plate.

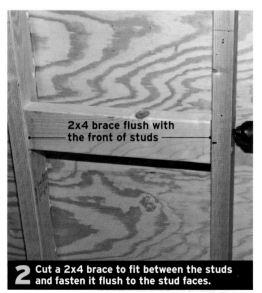

2x4 brace flush with the front of studs

2 Cut a 2x4 brace to fit between the studs and fasten it flush to the stud faces.

INSTALLING A PEDESTAL SINK (CONTINUED)

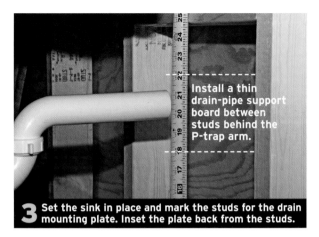

3 Set the sink in place and mark the studs for the drain mounting plate. Inset the plate back from the studs.

Install a thin drain-pipe support board between studs behind the P-trap arm.

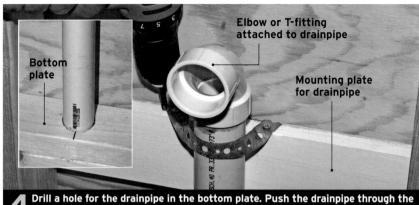

Elbow or T-fitting attached to drainpipe

Bottom plate

Mounting plate for drainpipe

4 Drill a hole for the drainpipe in the bottom plate. Push the drainpipe through the hole, and secure the pipe to the mounting plate. Then attach the drain elbow.

7 Install the finished wall and the sink support flange.

8 Remove the tape from pipe nipples and install stop valves.

P-trap arm

9 Install the P-trap on the tail piece and cut the trap arm to fit in trap adapter.

with its front face flush with the rear edge of the drain-pipe **3**. Then drill a hole in the floor plate for the drainpipe, insert the pipe into the hole, and secure it to the plate as shown **4**. Add an elbow or T-fitting to the drainpipe, depending on your design, and slide the sink up to the fitting to make sure it will line up correctly when installed.

➤ **See "Choosing Fittings," p. 60.**

Drill holes in the bottom floor plate for the supply lines, then attach the supply lines to drop-ear elbows, using the techniques and tools appropriate to the kind of pipe you're using **5**.

➤ **See "Supply Pipes," pp. 14–53.**

Fasten another support plate for the supply-line drop-ear elbows between the studs. Insert the supply lines through floor plates

and secure the elbows to the support plate **6**. Glue a trap adapter to the drain fitting, then thread finished pipe nipples on the drop-ear elbows. Tape the ends of the nipples. Then install the finished wall **7**, cutting holes for all pipes. Attach the sink bracket (metal or wood, depending on the manufacturer's specs), driving fasteners into the brace fastened between the studs.

Once the finished wall is up, install stop valves on the nipple ends (remove the tape) **8**. Then cut the end of the trap extension to fit, and clean off the burrs **9**. Install the faucet **10** and the supply tubes **11**, move the pedestal in place, and set the bowl on it **12**. Make sure the back of the sink catches the bracket on the wall. If necessary, adjust the P-trap arm to fit into the drain fitting, by moving the trap up or down on the sink tail piece. You may also have to move the

13 Hand-tighten the supply-tube nuts on the stop valves. Finish with a wrench.

pedestal slightly so the bowl sits securely. Tighten the trap adapter nut on the P trap arm. Attach the supply tubes to the stop valves **13**. Install the drain and pop-up assembly, and you're done **14**. Stand back to make sure everything looks right. Turn the stop valves on and check for leaks.

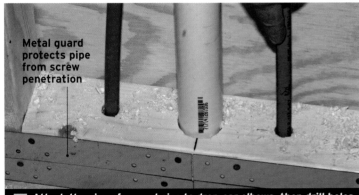

Metal guard protects pipe from screw penetration

5 Attach the pipe of your choice to drop-ear elbows, then drill holes in the floor plate for the pipes and insert the pipes in the holes.

6 Screw the drop ears to the support and install pipe nipples. Tape the nipples and glue a trap adapter to the fitting.

10 Fasten the faucet in the holes in the bowl before mounting the bowl.

11 Install the supply tubes on the faucet threads and tighten the nuts.

12 Place the bowl on the pedestal, fitting the P-trap arm into the trap adapter.

14 It's done. Turn water on at the stop valves and check for leaks.

USING THE RIGHT FITTINGS

Install the right drain fittings. Long sweep fittings (far right) can siphon the trap. Use an elbow (near right) for a ventless installation. Use a T-assembly if using a vent pipe.

A VESSEL-SINK CONVERSION

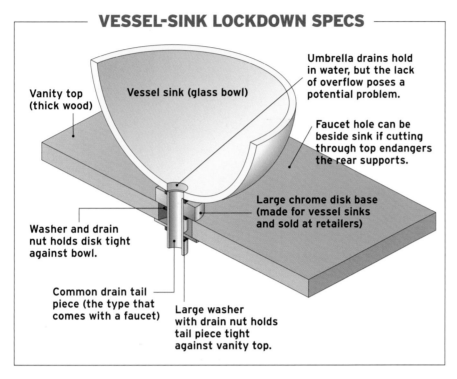

VESSEL-SINK LOCKDOWN SPECS

Vanity top (thick wood)

Vessel sink (glass bowl)

Umbrella drains hold in water, but the lack of overflow poses a potential problem.

Faucet hole can be beside sink if cutting through top endangers the rear supports.

Large chrome disk base (made for vessel sinks and sold at retailers)

Washer and drain nut holds disk tight against bowl.

Common drain tail piece (the type that comes with a faucet)

Large washer with drain nut holds tail piece tight against vanity top.

TYPICAL VANITY ROUGH-IN

Typically, the vanity drain is dead center (left to right), but it needs to be offset to the side by several inches if an AAV is to be used or if the drain will be in the way of a drawer.

32 in., typ.; but can vary

6 in., typ.

1½ in. PVC pipe terminating in a trap adapter (where the P-trap slides into pipe and tightens down with a large plastic nut)

Typically 16 in. but can vary from vanity to vanity

Can be even, above, or below drain line

Trim will terminate against side of vanity.

One of the newest trends in interior design is cabinetry that looks like furniture. You can cash in on this trend—and save yourself some cash—by picking an attractive piece of furniture and converting it into a vanity for a vessel sink.

Installing the sink

First, find a piece of furniture that matches the decor of the room. Mark the location of the bowl, typically dead center on the top of the piece. Make sure, however, that the bowl does not extend beyond the back of the furniture where it will bump into the wall ❶. Using a hole saw, drill a hole large enough to accommodate the tail piece and drain assembly ❷. (For large bottom concave sinks, cut the hole with a jigsaw.) Then with a homemade or manufacturer's template, drill holes for studs if your sink requires them. Attach the tail piece to the sink ❸. Then set the tail piece into the hole and note any parts of the vanity that will block the drain or water lines ❹. Drill through or remove anything that's in the way, but don't weaken the structure unless you can reinforce it later. If you make a drawer inoperable in this process, you can remove the drawer front and glue it back on the cabinet. Guided by the manufacturer's specs, mark the location for the faucet and drill its hole with a hole saw, typically just big enough for the supply tubes and the attachment rod. Insert the faucet and tighten it down ❺. Slide your new vanity into position, attach the P-trap and water lines, turn the water on and check for leaks.

➤ **For installation steps, see "Installing a Pedestal Sink," pp. 139–141.**

Installing the vanity

Although the shortest vanity will be around 19 in. long, the longest can go on as long as your pocketbook, but installation is the same. First rough in the plumbing if it's not already in place, using either the typical plan shown (at left) or a custom plan from the manufacturer. Then install the faucets and water lines on the sink and glue the sink to the frame. Last, screw the back into the wall and hook up the plumbing.

➤ **See "Installing Faucets," pp. 144–145.**

Sizing the sink recess. Open a hole large enough for the drain assembly. Some sinks will require additional holes for studs. Others will need a large area cut out for the bottom of the sink. Oval glass vessels will need metal bases to give them flat supporting surfaces. Make sure to install a high-tower faucet. They are several inches taller than standard faucets. (Right photo courtesy of Kohler)

1 Mark the location for the center of the vessel.

2 Measure the drain nut and drill a hole to accommodate it and any studs.

3 Fasten the tail piece to the sink, including the metal base for an oval glass vessel.

4 Remove any structural parts that are in the way of the drain tail piece. Install the sink and drill the hole for the high-tower faucet.

5 Install the high-tower faucet, slide the vanity into position, admire, then attach the water and drain lines.

INSTALLING A TWO-HANDLED BATH FAUCET

There are two basic types of bath faucets: two-handled models with threaded shanks and single-handled units with supply tubes attached. Most standard bath sinks are made with three holes to accommodate either style. The body of the threaded-shank design will cover all the holes, as will the base plates that come with single-handled models.

The easiest of all the styles to install are those with ½ in. shanks. Simply slide the faucet into the sink holes **❶**, and lock them tight to the sink body from below, using the nuts provided by the manufacturer **❷**. If the faucet doesn't come with a seal on its base, run a bead of silicone caulk along the perimeter of the faucet base before inserting it in the holes. Wipe off the excess when you're done. Then connect the water supply line to the stop valves.

➤ See "Installing a Pedestal Sink," pp. 139-141.

❶ Half-inch shank (4-in. center) bath faucets install fast and easy. Insert the faucet into the sink holes.

❷ Working underneath, secure the faucet by tightening the nuts on the shanks.

WIDESPREAD FAUCETS

Widespread faucets are the most expensive of faucet models. They don't work any better than any other faucet, but they can add an attractive dimension to your bath design scheme. They are also more difficult to install than other models because their hoses leave little room for wrenches in the tight cavity under the sink.

To install a widespread set, you will need a basin wrench, a small adjustable wrench, and various open-end wrenches to work in the tight space. You may even need some wood shims or braces to stop one thing from moving while you tighten another.

Start by inserting the shank of the hot-water valve through the left hole in the sink. Then tighten it with the nut provided. Repeat the process for the cold-water valve and the spout **❶**. Then interconnect the supply tubes to all three pieces that extend through the sink. Both hot- and cold-water lines connect to the spout with the supplied hoses **❷**.

❶ Start installing a widespread faucet by attaching the hot water faucet, then attach the spout and the cold water faucet.

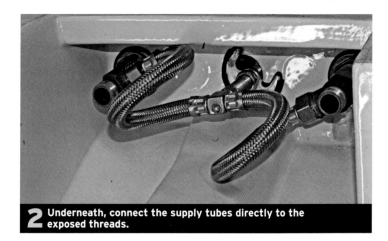

❷ Underneath, connect the supply tubes directly to the exposed threads.

INSTALLING A SINGLE-HANDLED BATH FAUCET

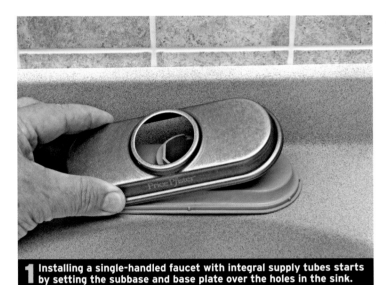

1 Installing a single-handled faucet with integral supply tubes starts by setting the subbase and base plate over the holes in the sink.

2 Insert the supply tubes (specially designed for the sink holes) one at a time through the central hole.

3 Once the supply tube nuts are through, slide the faucet to the base plate.

4 Slip the washer on the shaft and hand tighten the nut.

5 Lock the faucet securely by tightening the nut with an adjustable wrench.

Single-handled faucets come with a base plate that covers the side holes. The faucet body locks down from below by tightening a nut and washer on a threaded shaft that extends from the faucet body.

Set the plastic subbase in place, sealing it with silicone if necessary. Then set finished base plate on top of the subbase **1**. Insert the 1/2-in. male supply-line nuts one at a time into the center hole **2**, and then slide the faucet body into the hole and onto the base **3**. Working from below, slide the crescent-shaped washer on the threaded rod (holding it in place if necessary with silicone caulk) and tighten it down with the long-shank nut **4**. Hand tighten this nut first, then finish the job with an adjustable wrench **5**. Connect the water supply lines to the supply tubes.

TRADE SECRET
Installing any kind of faucet is easier done before you mount the sink. Never use plumber's putty to seal a faucet base, as it will always harden and crack over time. Instead, use clear silicone caulk.

INSTALLING A BATH SINK DRAIN

HOW A BATH DRAIN WORKS

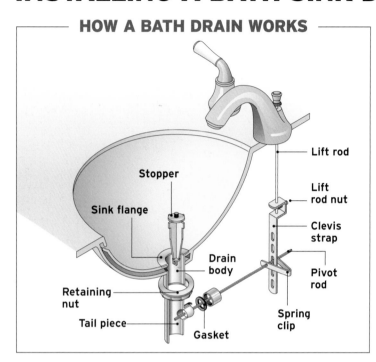

Lift rod

Lift rod nut

Clevis strap

Pivot rod

Spring clip

Stopper

Sink flange

Drain body

Retaining nut

Tail piece

Gasket

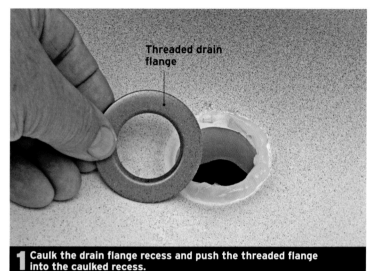

Threaded drain flange

1 Caulk the drain flange recess and push the threaded flange into the caulked recess.

4 Make sure the hole in the drain body faces to the rear.

5 Screw the tail piece to the drain.

When you're installing a new sink or faucet, you'll want the finish of the pop-up in the drain to match the finish of the faucet. So you'll want to put in a new drain, too. Although there are differences in drain design—some are plastic, others metal; some tail pieces screw into the drain, others come as one-piece units—installation is pretty much the same.

Start by caulking around the sink flange recess in the sink bowl with a silicone sealant and push the drain flange down into the caulked recess **1**. Caulk the underside of the drain and push the drain body up into the flange **2**. Hold the drain flange in its recess with one hand and with the other, screw the drain body into the flange with your fingers, then snug it with an adjustable wrench **3**. Make sure the hole for the pop-up linkage in the drain body faces to the rear **4**. Put silicone caulk on the threads of the tail piece, and screw it on the drain body **5**. Insert the lift rod in the hole in the faucet body, making sure that the concave seal inside the hole is in place **6**. Assemble the clevis strip extension to the lift rod, tightening the clevis nut slightly for now **7**. Drop the pop-up stopper into the drain with the hole in its lower tab facing to the rear, then slide the pop-up pivot rod into the hole in the faucet body and through the stopper tab **8**. Slide on the gasket, if there is one, then tighten the retaining nut with pliers or small wrench **9**. Insert the pivot rod into the clevis extension and snug it with the spring clip **10**. Test the stopper. When you push down on the lift rod, the stopper should rise. Pulling up on the rod should bring the stopper down. Adjust the position of the pivot rod, the pop-up rod, and clevis strap until the stopper seals the drain when the rod is up, but lets sufficient water flow when the rod is down.

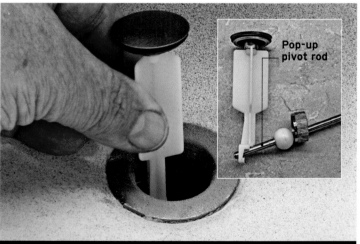

Pop-up pivot rod

8 Insert the pop-up stopper into the drain with tab facing to the rear so it will accept the pop-up pivot rod.

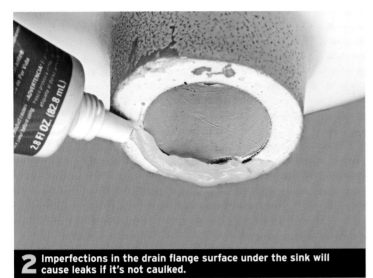

2 Imperfections in the drain flange surface under the sink will cause leaks if it's not caulked.

3 Thread the drain body into the flange and use a wrench to tighten it.

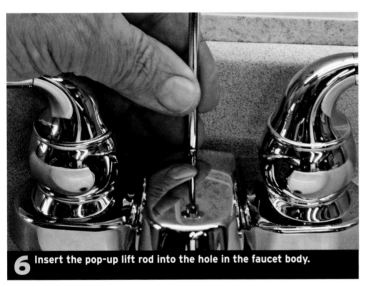

6 Insert the pop-up lift rod into the hole in the faucet body.

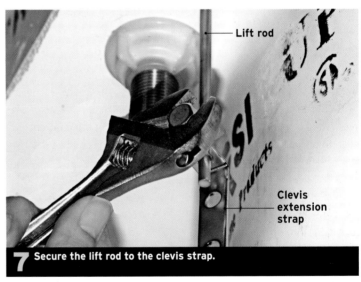

Lift rod

Clevis extension strap

7 Secure the lift rod to the clevis strap.

9 Insert the pivot rod into drain and tighten the retaining nut.

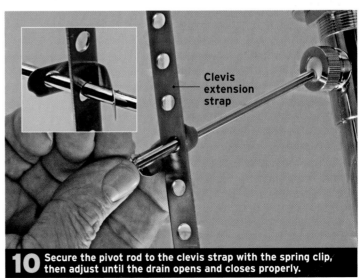

Clevis extension strap

10 Secure the pivot rod to the clevis strap with the spring clip, then adjust until the drain opens and closes properly.

INSTALLING A TUB OR SHOWER

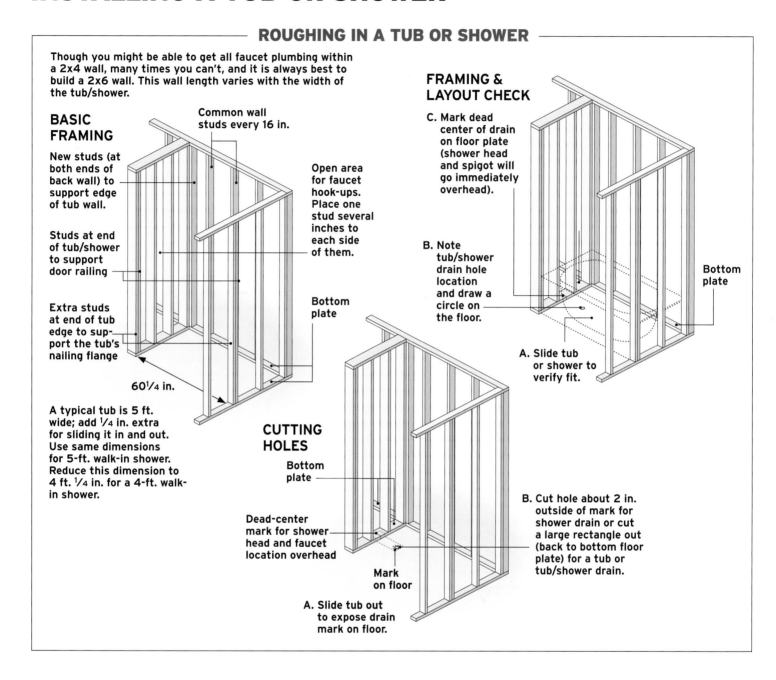

ROUGHING IN A TUB OR SHOWER

Though you might be able to get all faucet plumbing within a 2x4 wall, many times you can't, and it is always best to build a 2x6 wall. This wall length varies with the width of the tub/shower.

BASIC FRAMING

Common wall studs every 16 in.

New studs (at both ends of back wall) to support edge of tub wall.

Studs at end of tub/shower to support door railing

Extra studs at end of tub edge to support the tub's nailing flange

Open area for faucet hook-ups. Place one stud several inches to each side of them.

Bottom plate

60¼ in.

A typical tub is 5 ft. wide; add ¼ in. extra for sliding it in and out. Use same dimensions for 5-ft. walk-in shower. Reduce this dimension to 4 ft. ¼ in. for a 4-ft. walk-in shower.

FRAMING & LAYOUT CHECK

C. Mark dead center of drain on floor plate (shower head and spigot will go immediately overhead).

B. Note tub/shower drain hole location and draw a circle on the floor.

A. Slide tub or shower to verify fit.

Bottom plate

CUTTING HOLES

Bottom plate

Dead-center mark for shower head and faucet location overhead

Mark on floor

A. Slide tub out to expose drain mark on floor.

B. Cut hole about 2 in. outside of mark for shower drain or cut a large rectangle out (back to bottom floor plate) for a tub or tub/shower drain.

The basics for installing a tub, shower, and even a whirlpool are all the same. Units for new construction must be installed during the rough-in stage before the finished walls are up, preferably before the bathroom door framing goes in. In both old and new construction, you can install a new tub or shower against any kind of finished wall.

Typically, a bath or shower unit consists of some type of base and a surround if there is a shower. Showers and tub/showers can be installed as one unit if you have room to get them through the existing doorway. If you're installing this kind of unit in an existing bath, purchase the base and surround as separate items. Before you purchase a unit, study the layout illustrations shown above.

Build the walls first. Mark the floor for any bottom plates you need to install, then hang studs between a top and bottom plate. Note the number of studs you'll need. On both end walls, install at least four studs for a tub/shower unit; one for the corner, one in the middle, one for the door-mounting rail, and one for the outer edge ❶. Slide the tub in place and mark the outline of the drain hole on the subfloor ❷. Remove the tub and cut a section out of the floor (typically 8 in. by 10 in. or larger) for the tub drain.

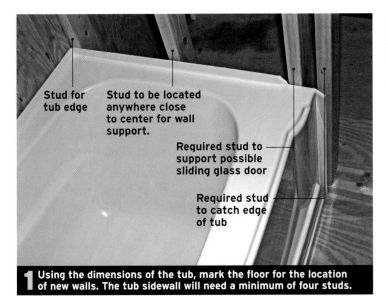

Stud for tub edge

Stud to be located anywhere close to center for wall support.

Required stud to support possible sliding glass door

Required stud to catch edge of tub

1 Using the dimensions of the tub, mark the floor for the location of new walls. The tub sidewall will need a minimum of four studs.

2 Insert the tub in its framed enclosure and draw a circle for the location of the tub drain.

3 Level the short sides of the tub, if necessary.

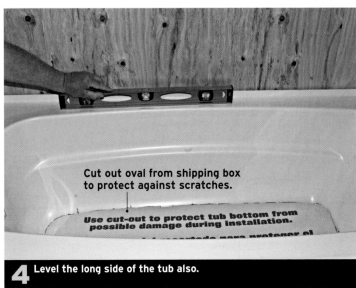

Cut out oval from shipping box to protect against scratches.

Use cut-out to protect tub bottom from possible damage during installation.

4 Level the long side of the tub also.

➤ **See Illustration, "Roughing in a Tub or Shower," on the facing page.**

Slide the unit in place and level it **3**. The level does not have to be perfect but sufficient for the water to drain **4**. Once leveled, shoot a furring strip on the wall (a solid wall

» » »

REMOVING AN OLD TUB

To remove an old tub, first disconnect all plumbing, and then rip out the finished wall surface over and around the tub about 6 in. Steel tubs can be picked up and carried out; cast iron will have to be broken up and taken out in pieces. You will need a sledge hammer, a circular saw with a diamond blade, lots of prybars, and perhaps a hydraulic jack. The work is slow and dangerous, but can be done. Wear gloves and safety glasses. Cover all windows with plywood, prop plywood around the toilet and vanity to protect them—large razor-sharp shards will fly everywhere. Once you get the large pieces out, vacuum up the shards.

INSTALLING A TUB OR SHOWER (CONTINUED)

or studs) just under the tub back (long side) **5**. If the tub body material is thin, you will need another furring strip behind the front tub wall **6**. Predrill the nailing flange at each stud location. You will fasten the lip later **7**. Pull the tub back out and fasten the drain to the tub.

➤ See "Installing a Drain in a New Tub," p. 152.

To fill the $\frac{1}{4}$-in. space between the flange and the wall, shim the wall out with a $\frac{1}{4}$-in. Hardiboard® panel fastened to the studs or the wall surface. Predrill the Hardiboard and drive in drywall screws **8**. Then set the tub in place, slipping the drain into the hole you've cut. Fasten the tub flange to the studs with flat-head roofing nails. Following the manufacturer's instructions, glue the

surround panels directly to the Hardiboard, resting the bottom of the surround panels on paint-stirring sticks to keep the panels $\frac{1}{8}$ in. above the tub shelf **9**. When all the surround panels are up and the glue has set, remove the spacers and caulk the joint. Then hook up the drain and supply lines.

5 Along the back wall below tub, fasten a furring strip for support.

6 Fasten a second furring strip to the floor behind the tub skirt.

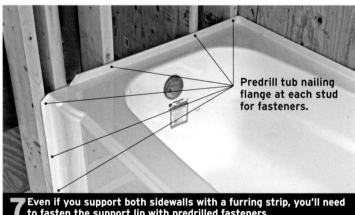

Predrill tub nailing flange at each stud for fasteners.

7 Even if you support both sidewalls with a furring strip, you'll need to fasten the support lip with predrilled fasteners.

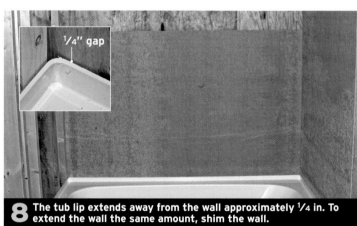

1/4" gap

8 The tub lip extends away from the wall approximately $\frac{1}{4}$ in. To extend the wall the same amount, shim the wall.

Behind the bath or shower Drywall behind a tub or shower will rot and so will Greenboard™. Hardiboard or cement backer board will not. Both are sold everywhere. The weight of cement board prohibits it being made in 5-ft. widths; the width of a tub or tub/shower. Hardiboard is lighter and comes in 5-ft. widths. When cut, both will make quite a mess. Cut outdoors in an open area downwind from the saw.

9 Glue the surround to the walls, following the manufacturer's instructions. Install the plumbing and fixtures and check for leaks.

TUB AND SHOWER SURROUNDS

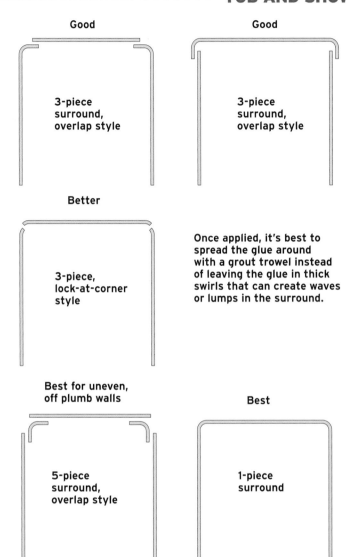

Good

3-piece surround, overlap style

Good

3-piece surround, overlap style

Better

3-piece, lock-at-corner style

Once applied, it's best to spread the glue around with a grout trowel instead of leaving the glue in thick swirls that can create waves or lumps in the surround.

Best for uneven, off plumb walls

5-piece surround, overlap style

Best

1-piece surround

Types of surrounds

Different types of surrounds are meant for different installations. The best surround (assuming the limits of your space prohibit an integral tub and shower and surround) is a 1-piece surround. The fewer pieces you have to put together, the faster the install and the less chance of a leaky seam. However, a 1-piece surround needs straight walls and a flat base behind it, and you might not be able to get it in through the bathroom door. The 3-piece lock-at-the-corner style functions like a single unit once assembled, and you can get the individual pieces through doors. Both styles assume good carpentry practices. For older houses with bumpy, wavy walls and a tub base that is not plumb, you need to have the side walls separate from the back since the walls around the tub may not be at right angles. Surrounds come in various thicknesses—typically, the thicker the better (and more expensive). However, thin surrounds have a place in old houses because they can better follow the wavy walls.

Basic surround rules

Regardless of type, all surrounds must be installed properly. It is the lip or flange, not the surround itself, that gets nailed. Never hammer a nail directly into a surround lip—it will split and shatter. Always predrill a hole for the fastener.

   To attach a flanged surround, glue it and nail it. If the surround has no lip or edge, use glue only. Heavy-duty surrounds can be installed on any surface—on a solid wall or on studs—but very thin surrounds will need a solid surface all round. Use Hardiboard or cement backer board.

   The practice of installing all studs 16 in. on center doesn't always apply to tub and shower installations because you will need extra studs for the surround corners and the door. In addition, you cannot place a stud where the faucet will be installed (right above the drain). Remove any existing stud in this location and install new studs on each side of the faucet.

Whirlpool tubs

Installation of a whirlpool tub is the same as for a tub or tub shower, except whirlpool tubs may be physically longer. You must locate the pump, however, in an accessible area because you'll need to maintain it. You will also have to get power to both it and a heater if the tub has one. Fancier whirlpools will require a couple of mini walls the height of the tub on two sides. Some will feature a finished face (painted drywall or tile). Others will come with a plastic surround—but behind the surround, you'll need to build a mini stud wall.

This whirlpool requires the creation of a finished wall on the two sides facing out. (Photos courtesy Kohler)

This whirlpool has a surround and won't need a finished wall, but you'll have to build a mini stud wall behind the tub.

INSTALLING A DRAIN IN A NEW TUB

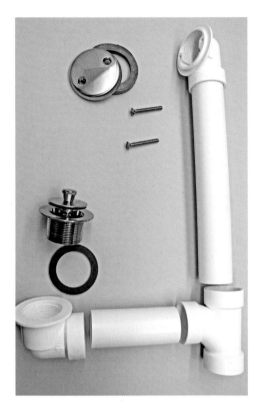

A tub pull-up drain is easy to install and maintain. Simply pull up the disk to drain the water.

1 Unscrew the center pop-up from the drain body.

2 Caulk the underside of the drain flange and the recess in the tub.

CABLE-CONTROLLED DRAIN

On a cable-controlled drain, turning the handle pushes or pulls the cable, which raises or lowers the stopper.

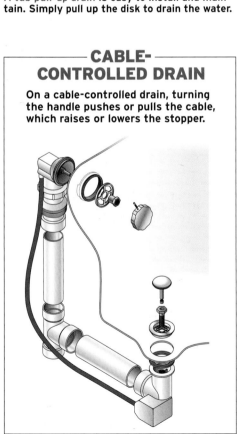

6 Fasten the overflow cover to the tub body.

7 Assemble the pipe between the overflow and the drain.

Modern tub drains are better than ever. Gone are the days of complicated linkages. Today's drains are simple: both pull-up drains and their cabled counterparts.

Both drain styles use common schedule 40 PVC drainpipe and glued fittings. These models install a lot faster than older designs and make rodding the pipes through the tub overflow easier. The pull-up drain combines simplicity and low cost; it also drains fast. For a fancier drain that doesn't get your hands wet when pulling the plug in a full bath, opt for a cable-drain mechanism.

Installing a pop-up tub drain

Remove the center pop-up by unscrewing it from the drain **1**. Caulk both the underside of the drain flange and the recess in the tub body into which it fits **2**. Insert the drain body into the tub hole **3**. Insert the open jaws of needle-nose pliers in the drain ports into the drain (to keep it from turning) **4** and

3 Fit the drain body in the hole in the bottom of the tub.

4 Insert needle-nose pliers in the drain so it won't turn when you tighten the elbow.

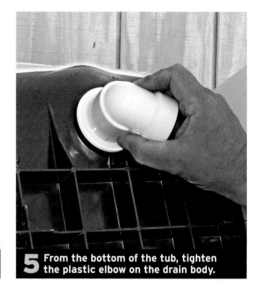

5 From the bottom of the tub, tighten the plastic elbow on the drain body.

8 Hold a 3-in. Femco Flex T next to the drain and mark cut lines on the pipe.

9 Cut the drainpipe on your marks with a reciprocating saw.

10 Compress the T into the gap and tighten the clamps.

tighten the plastic drain elbow on the threads of the drain body **5**. Fasten the overflow cover to the tub **6**, and dry-fit the drainpipe between the overflow and the tub drain **7**. If the PVC pipe supplied with the kit does not fit perfectly between the overflow and the drain fittings, cut it to fit. Once all pieces fit, glue the joints and install the tub.

➔ See "Installing a Tub or Shower," pp. 148–151.

Use a flexible T-fitting to connect the tub drain to the main drainpipe. Hold the T-fitting against the drainpipe to mark its dimensions. Be sure to allow for the flanges on the fitting **8**. Then cut the drainpipe on your marks **9**. Insert the fitting into the gap in the drainpipe, point it to receive the tub drain, and tighten the clamps **10**. Insert the trap into the T-fitting, using a reducing bushing (if necessary) and attach the tub drain to the fitting. Then tighten all clamps **11**.

11 Insert the tub-drain trap pipes into the T and tighten all clamps.

INSTALLING A SINGLE-HANDLED SHOWER FAUCET

A single-handled shower faucet will typically need a very large hole cut for it, and to cut it, you'll need a large-diameter hole saw (preferred), a jigsaw, or a reciprocating saw. If you're hanging a shower surround, cut the hole in the finish surround and set it aside. If you're installing tile, mount the faucet body first, then cut the hole in the finished wall. You can either measure the diameter of the faucet body or use a template, if supplied by the manufacturer. Using the manufacturer's spec sheet, locate and mark the exact center of the faucet on the finished wall surface, typically in line with the drain and shower head and about 40 in. above the shower floor. Remove any stud behind this location and reinstall it where it won't be in the way ❶. Then cut the hole and dry-fit the escutcheon plate to test its fit ❷. Enlarge the hole with a rasp if necessary, then set the wall panel aside. Note how your faucet is configured for mounting, this varies considerably among models. The faucet shown here comes with its own mounting brackets ❸. Others may not. You may want to create your own mounting system by gluing boards around the open hole to secure the pipe, which in turn secures the faucet. Assemble the water lines to the faucet body, install the necessary framing, and mount the faucet body ❹.

Mount the escutcheon over the faucet body with the supplied screws, making sure that the embossed designations for hot and cold water are in their proper places ❺. Attach the handle with the supplied screw. Turn the handle to make sure it operates smoothly, then hook up the water lines.

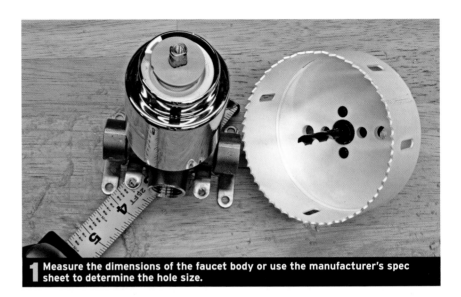

1 Measure the dimensions of the faucet body or use the manufacturer's spec sheet to determine the hole size.

2 Using the proper diameter hole saw, cut the opening through the finished wall.

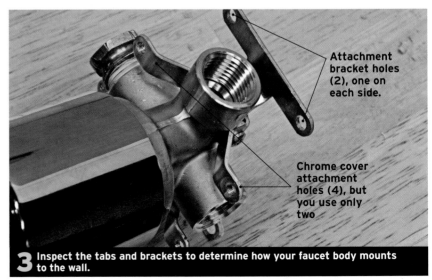

Attachment bracket holes (2), one on each side.

Chrome cover attachment holes (4), but you use only two

3 Inspect the tabs and brackets to determine how your faucet body mounts to the wall.

TRADE SECRET

Whenever you work with a paste pipe compound, have a roll of paper towels around to constantly clean your hands. Otherwise you will get the stuff on your clothes and ruin them.

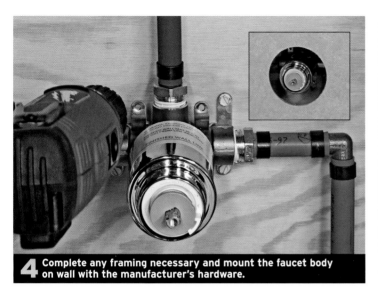

4 Complete any framing necessary and mount the faucet body on wall with the manufacturer's hardware.

5 Orient the escutcheon so the lettering is in the proper position, mount the escutcheon and attach the handle.

SINGLE-HANDLED SHOWER BODIES AND HANDLES

Shower bodies can come with either male or female threads. Avoid copper sweat fittings at all costs. Sweating copper can often destroy something on the inside of the faucet.

Look for thick solid-metal handles held on the faucet with a Phillips screw passing through the center of the handle into the center of the stem. Avoid handles with Allen-head screws threaded in their sides. When you have to remove the handle, you won't remember what size tool to use and whether it was metric or not. There will always be a Phillips-head screwdriver around. Even if you do remember, odds are you'll strip the screw and will have to break the handle to get it off.

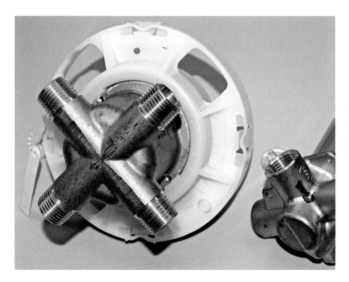

Make sure the showerbody **has either male or female threads. Avoid copper sweat fittings.**

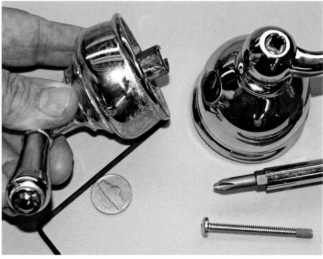

Avoid set-screw attachments (left). Always opt for the old-fashioned Phillips-head screw inserted through the top of the handle.

INSTALLING A TWO-HANDLED TUB/SHOWER FAUCET

ASSEMBLY OF A TUB/SHOWER FAUCET

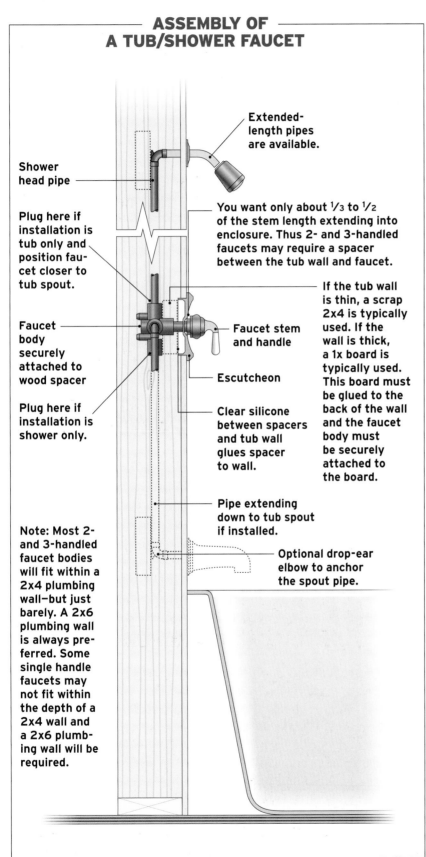

Extended-length pipes are available.

Shower head pipe

Plug here if installation is tub only and position faucet closer to tub spout.

You want only about 1/3 to 1/2 of the stem length extending into enclosure. Thus 2- and 3-handled faucets may require a spacer between the tub wall and faucet.

If the tub wall is thin, a scrap 2x4 is typically used. If the wall is thick, a 1x board is typically used. This board must be glued to the back of the wall and the faucet body must be securely attached to the board.

Faucet stem and handle

Faucet body securely attached to wood spacer

Escutcheon

Plug here if installation is shower only.

Clear silicone between spacers and tub wall glues spacer to wall.

Pipe extending down to tub spout if installed.

Optional drop-ear elbow to anchor the spout pipe.

Note: Most 2- and 3-handled faucet bodies will fit within a 2x4 plumbing wall—but just barely. A 2x6 plumbing wall is always preferred. Some single handle faucets may not fit within the depth of a 2x4 wall and a 2x6 plumbing wall will be required.

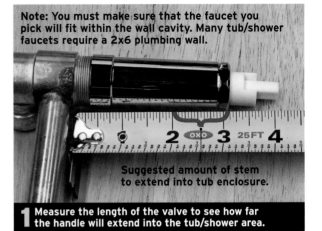

Note: You must make sure that the faucet you pick will fit within the wall cavity. Many tub/shower faucets require a 2x6 plumbing wall.

Suggested amount of stem to extend into tub enclosure.

1 Measure the length of the valve to see how far the handle will extend into the tub/shower area.

Most 2-handled faucets use long cylinders to regulate the water. While they will fit front-to-back in a 2x6 wall, in a 2x4 wall, their length extends dangerously past the interior wall surface. As a remedy, glue a wood spacer (a 2x4 or 1x4, depending on the thickness of the wall). This will leave just enough of the faucet arm extending into the stall to mount the handle.

To start, measure the arm and the wall it penetrates to compute the thickness of the wood spacer, here a 2x4 **1**. Cut the spacer slightly longer than the width of the valve centers. Then cut recesses in the ends to accommodate the valves **2**. Secure the faucet to the spacer with common copper clamps. Install a plug in the bottom of the output fitting if you're not going to install a tub spout **3**. For a tub/shower, leave it open for the spout pipe.

➡ **See "Which Side Up?," p. 158.**

In the top of the output fitting, install a male adaptor appropriate to the pipe you're using to feed the shower head (here, PEX) and hook up the supply lines to the faucet body **4**. Hold the tab of your tape on the outside of one valve stem and measure to the inside of the other. This will give you the center-to-center spacing, here 8 in. Mark the wall for this spacing **5**. To determine the diameter of the hole, cut a hole in cardboard with your hole saw. Fit the cardboard over the valve body to make sure you have the correct size hole saw **6**, then drill two holes in the tub wall at your marks **7**. Insert the faucet into the holes, gluing the spacer to the rear of the finished surround wall **8**. Then slip the escutcheons over the handle arms and snug them against the wall. Once you're sure everything fits, pull the escutcheons out, caulk behind them, and then push them back in place. Install the handles, and hook up the water lines **9**.

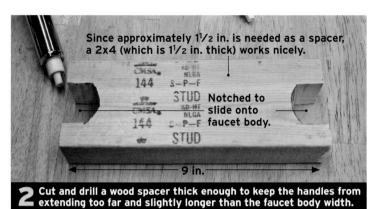

Since approximately 1¹/₂ in. is needed as a spacer, a 2x4 (which is 1¹/₂ in. thick) works nicely.

Notched to slide onto faucet body.

144 STUD
144 STUD

9 in.

2 Cut and drill a wood spacer thick enough to keep the handles from extending too far and slightly longer than the faucet body width.

3 Fasten the faucet to the spacer with copper straps. If you're not going to have a tub spout, install a ¹/₂-in. plug in the output.

Male adaptor for pipe to shower

4 Install a male adaptor for your shower pipe in the shower outlet and hook up the supply lines.

5 Measure from the outside of one handle stem to inside the other, typically 8 in., and transfer the measurement to the wall.

6 Measure the size of the valve and make a cardboard template with a hole saw to check its fit, ¹/₄ in. in diameter larger than the arm.

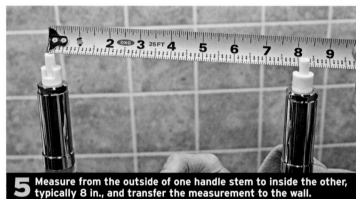

7 At the marks you made in step 5, drill two holes in shower wall for the faucet handle arms.

8 Apply silicone between the spacer and the tub wall, insert the faucet arms, and press the assembly in place.

9 As always, the last thing to do is to install the escutcheons and handles.

INSTALLING A THREE-HANDLED TUB/SHOWER FAUCET

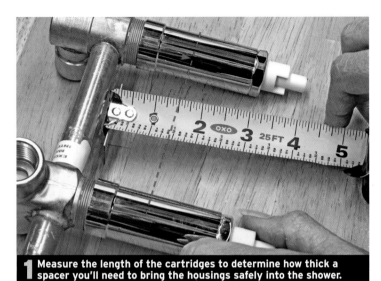

1 Measure the length of the cartridges to determine how thick a spacer you'll need to bring the housings safely into the shower.

2 Cut the spacer slightly wider than the faucet, and drill holes for the arms and diverter valve. Attach with copper clamps.

A three-handled tub/shower faucet can lend a nostalgic design element to your bath. Like two-handled models, these faucets need a spacer on the rear of the finished wall to keep the handles from protruding excessively into the tub or shower stall. The thickness of the spacer will depend on the interior space in the wall.

A three-handled unit requires essentially the same steps as a two-handled faucet. Measure the length of the valve bodies ❶; cut a spacer of the right thickness and fit it with holes for the three-handled unit ❷; fasten the faucet body to the spacer; insert plugs or adaptors; depending on whether you're using the tub, shower, or both ❸.

➡ See "Installing a Two-Handled Tub/Shower Faucet," pp. 156–157.

Some faucets come with a plug for shower-only installations, others do not. If it's not supplied and you need one, get a 1/2-in. brass or galvanized plug, and screw it in the output that goes to the spout. For tub/shower installations, install male adaptors made for the water-supply pipe you're using, PEX in this case. Typically, you'll drill the wall at 0 in., 4 in., and 8 in. to accommodate the three elements that stick through the wall ❹,❺. Remember to make the holes 1/4 in. larger than the valve diameters ❻. Slide the faucet in, using silicone rubber to glue the spacer to the back of the wall, attach the escutcheons ❼, install the handles, and connect the water supplies ❽.

➡ See photo 7 on p. 162.

WHICH SIDE UP?

When installing any tub and shower faucet, you have to be sure you get the right side up. There are four fitting holes arranged in cross shape—two for the incoming water, one for the shower head and another for the tub spout. Fortunately, all you have to do is to look on the faucet—it will tell you. In the photo at right, note that the tub output is different than the shower output—they cannot be exchanged. Point the shower output up.

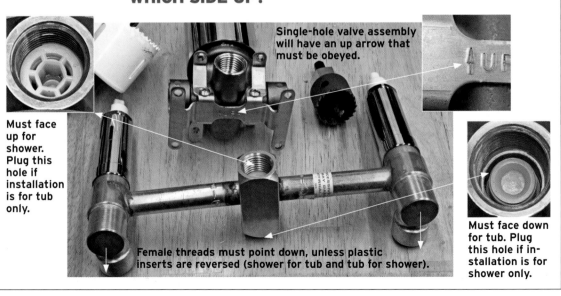

Single-hole valve assembly will have an up arrow that must be obeyed.

Must face up for shower. Plug this hole if installation is for tub only.

Female threads must point down, unless plastic inserts are reversed (shower for tub and tub for shower).

Must face down for tub. Plug this hole if installation is for shower only.

3 For showers only, as in this example, screw a plug in the tub-spout output and an interface fitting for the pipe going to the shower.

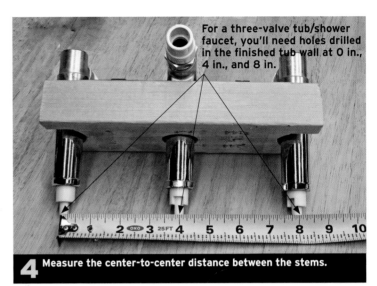

For a three-valve tub/shower faucet, you'll need holes drilled in the finished tub wall at 0 in., 4 in., and 8 in.

4 Measure the center-to-center distance between the stems.

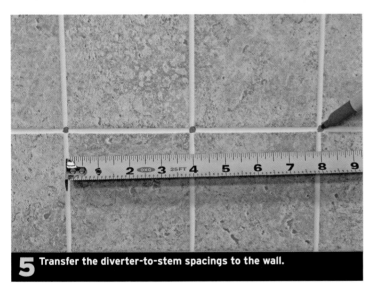

5 Transfer the diverter-to-stem spacings to the wall.

6 Using a hole saw ¼-in. larger than the valve diameters, drill holes in the finished wall.

7 Insert the stems through the wall, gluing the spacer to its rear surface and slide on the escutcheons.

8 Attach the handles and connect the water supplies.

REPLACING A TUB/SHOWER FAUCET

Replacing a tub/shower faucet means working in the wall behind the shower and cutting all the attached pipes to remove the unit. Have a bowl ready to catch any residual water in the pipes. Cut PEX or plastic piping with a ratcheting scissors. Use a hacksaw or a reciprocating saw with a fine-tooth blade to cut metal pipes and remove the unit ❶,❷.

Because the insets for the shower and tub are different, make sure to install the shower port pointing up. Plug the bottom port of the new body if you're not going to use it to supply a tub. If you need to support the unit on its lower side, thread in a 4-in. galvanized nipple in the bottom port and plug the nipple.

➤ See "Which Side Up?," p. 158.

Push-on street elbows that screw directly into the valve body are the fastest way to reconnect with the existing water lines. Wrap Teflon tape on the street elbow threads and screw the fittings all the way in. If you don't use push-on fittings, install threaded adaptors that will interface with whatever pipe you're using and glue, crimp, or solder the joints as appropriate.

➤ See "Supply Pipes," pp. 14-53.

Connect the shut-off valves to the supply pipes and the house pipes to the valves. Push-on shut-off valves, if used, will accept CPVC, copper, or PEX. Then push the new spigot body into the old location, clamping whatever pipes will secure the body. Here, we clamped the shower pipe and the bottom plugged nipple ❸.

The new valve should extend past the finished wall just enough to let you attach the handle and allow it to operate freely ❹. Fasten the cover plate to the wall, and attach the handles. If you haven't done so already, splice the new vertical pipe coming from the valve to the old shower head pipe.

1 Cut the pipes free from the rear.

2 If there are any holding clamps, remove them, and pull the spigot body from wall.

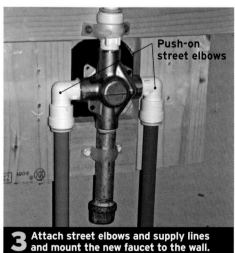

Push-on street elbows

3 Attach street elbows and supply lines and mount the new faucet to the wall.

4 Verify that the proper amount of body extends into the shower stall.

At the bottom of the foam cover plate **seal** (visible from the back) is a space to allow water to drain. Make sure to install the plate with this opening down.

ROUGHING IN A SHOWER HEAD

Before you install a shower head in new construction, you have to rough in the framing. That means installing support framing, piping, and a temporary nipple for the shower head.

Start by marking the height of the shower head on the studs. Cut a support cleat to fit the stud cavity, and fasten it to the studs, centered on your marks ❶. Attach a length of pipe (PEX is the easiest) to a drop-ear elbow, long enough to reach the faucet below. Screw the drop-ear elbow to the support board slightly above the final location of the shower head ❷. Although this location is typically about 7 ft. above the shower floor, there is no hard and fast specification for this height. Locate it to suit the user. Screw a 4-in. or 6-in. nipple into the drop-ear elbow, and tape the end of the nipple ❸. This pipe will take all the scratches and bangs it might get as you hang the finished wall. Once the finished wall is up, withdraw the nipple, and screw in finished shower head pipe. Use the end of pliers for extra torque if needed ❹,❺. Slip on the escutcheon, and screw on your shower head ❻. Then attach the water line.

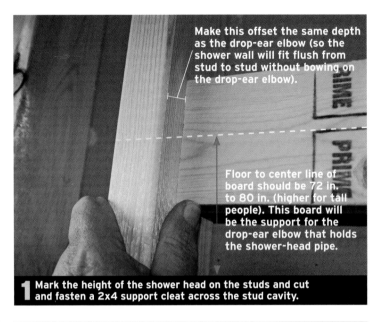

Make this offset the same depth as the drop-ear elbow (so the shower wall will fit flush from stud to stud without bowing on the drop-ear elbow).

Floor to center line of board should be 72 in. to 80 in. (higher for tall people). This board will be the support for the drop-ear elbow that holds the shower-head pipe.

1 Mark the height of the shower head on the studs and cut and fasten a 2x4 support cleat across the stud cavity.

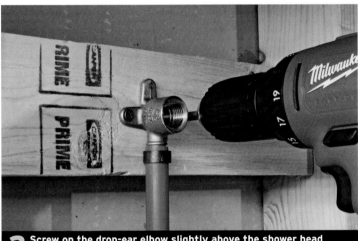

2 Screw on the drop-ear elbow slightly above the shower head location, typically around 7 ft. above the shower floor.

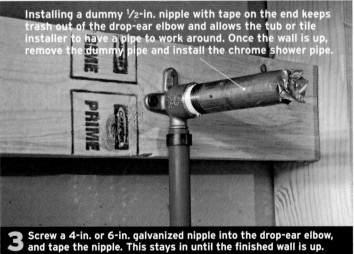

Installing a dummy ½-in. nipple with tape on the end keeps trash out of the drop-ear elbow and allows the tub or tile installer to have a pipe to work around. Once the wall is up, remove the dummy pipe and install the chrome shower pipe.

3 Screw a 4-in. or 6-in. galvanized nipple into the drop-ear elbow, and tape the nipple. This stays in until the finished wall is up.

4 Drill a hole for the nipple in the finished wall and hang the wall.

5 Untape the nipple and withdraw it. Install the finished shower-head pipe.

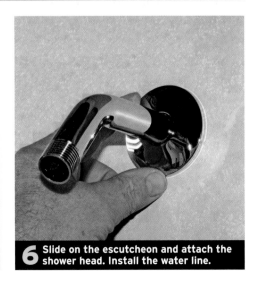

6 Slide on the escutcheon and attach the shower head. Install the water line.

MOUNTING A TUB SPOUT

There are several different types of tub spouts. The most common design comes with female 1/2-in. threads that accept a 1/2-in. nipple. A more modern type is made to slip on a copper nipple extending from the faucet body through the wall. Even newer is a universal spout made to hook up to just about anything. The copper slip-on is the easiest to install but is prone to pulling off if children or elderly individuals use it as a handle to exit the tub or catch their balance. The threaded model, as shown here, is the most secure ❶.

Because not all 6-in. galvanized nipples are exactly 6 in., your spout location will have to be tailor made. To start, install the galvanized 6-in. nipple in the tub port (bottom) of the faucet, then install the galvanized elbow. Use a pipe vise to make the job easier.

➤ See "Working with Cast-Iron and Galvanized Pipe," p. 71.

Measure from the center of the faucet to the center of the elbow outlet to determine the location of the hole you'll drill through the wall for the outlet nipple ❷. The distance from spigot to spout is variable, but experience has shown it needs to be around 2 in. above the spillover edge. If it's lower, the spout could be in the tub water; higher is fine, but too high will cause the water to splatter.

Install the faucet body if you haven't done so already.

➤ See "Installing a Three-Handled Tub/Shower Faucet," p. 158.

Then drill a hole for the spout at the vertical centerline of the faucet body ❸. It should fall exactly in line with the center of the elbow that will receive the spout nipple, just above tub spillover ledge by about 2 in. Insert a galvanized nipple of the required length for your spout into the galvanized elbow ❹, then screw the spout on ❺. The rear edge of the spout should just touch the wall (or be very close), but rarely do you get the distance right the first time. If you're more than a 1/2 in. off, change nipples ❻. Once you get close, add a shim on the rear of the wall to pull the spout tighter to the wall. A slight gap is tolerable, as long as you don't mind seeing some white or clear caulk. You can also buy a plastic circular-molded shim that is made to be semi-attractive and is designed to fill the gap between the spout and wall. ❼. Once you obtain a perfect fit, or one that you can live with, remove the spout, caulk around the hole, and screw the spout back on ❽.

WHAT CAN GO WRONG

You can, of course, use a copper nipple from the elbow to the spout, and copper will allow you to cut the nipple to the right length. But this convenience is offset by the tendency of copper-mounted spouts to pull off.

Universal tub spouts. They fit 1/2-in. threaded, 3/4-in. threaded, and copper pipe.

Common tub spout with 1/2-in. threads in its tip

2 in.

6-in. galvanized nipple

1 Pick a spout design. Copper slip-on models install quickly and more precisely, but spouts threaded for a galvanized nipple won't pull off.

Pipe from faucet body to spout must be rigid such as galvanized or copper. Length must allow spout to fit snug against wall.

4 Insert a galvanized nipple of the required length through the hole into the elbow behind the wall.

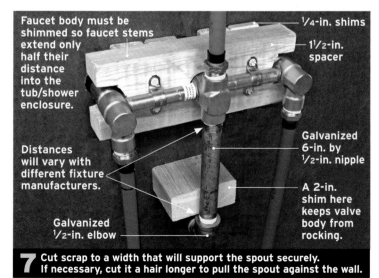

Faucet body must be shimmed so faucet stems extend only half their distance into the tub/shower enclosure.

1/4-in. shims

1 1/2-in. spacer

Distances will vary with different fixture manufacturers.

Galvanized 6-in. by 1/2-in. nipple

Galvanized 1/2-in. elbow

A 2-in. shim here keeps valve body from rocking.

7 Cut scrap to a width that will support the spout securely. If necessary, cut it a hair longer to pull the spout against the wall.

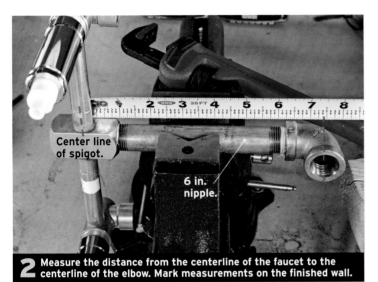

2 Measure the distance from the centerline of the faucet to the centerline of the elbow. Mark measurements on the finished wall.

Center line of spigot.

6 in. nipple.

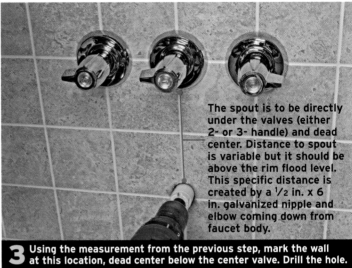

The spout is to be directly under the valves (either 2- or 3- handle) and dead center. Distance to spout is variable but it should be above the rim flood level. This specific distance is created by a 1/2 in. x 6 in. galvanized nipple and elbow coming down from faucet body.

3 Using the measurement from the previous step, mark the wall at this location, dead center below the center valve. Drill the hole.

5 Screw on the spout as a test.

6 If the space between the rear of the spout and the wall is 1/2 in. or more, insert a shorter nipple.

8 Caulk around the hole, and reinstall the spout.

⚠ WARNING

The pipe from the lower elbow to the spout should always be copper or galvanized. Never, under any circumstance, use plastic pipe for the spout pipe. It is easily bendable and breakable.

ROUGHING IN A TOILET FLANGE

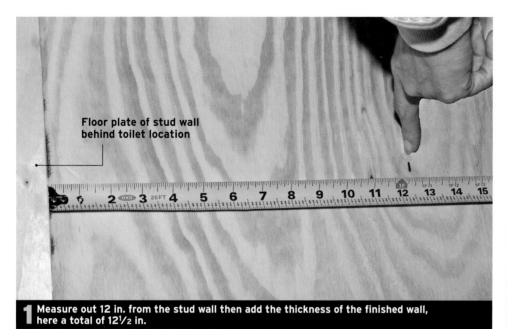

Floor plate of stud wall
behind toilet location

1 Measure out 12 in. from the stud wall then add the thickness of the finished wall, here a total of 12½ in.

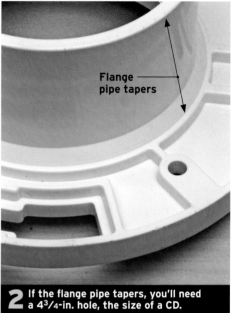

Flange
pipe tapers

2 If the flange pipe tapers, you'll need a 4¾-in. hole, the size of a CD.

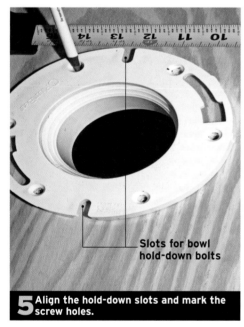

Slots for bowl
hold-down bolts

5 Align the hold-down slots and mark the screw holes.

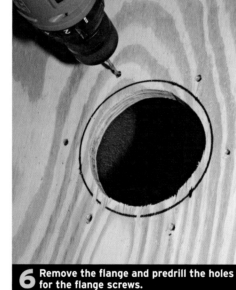

6 Remove the flange and predrill the holes for the flange screws.

7 Drive heavy duty flat-head screws into the floor to secure the flange.

Roughing in a toilet flange involves measuring for its location from the finished wall, cutting the hole for the toilet flange, and fastening the flange to the floor. The standard location of the center of the flange is a minimum of 12 in. from any finished wall. In addition, the bottom of the flange will have to be level with the finished floor. If you plan to lay tile, shim the flange up by the combined thickness of the tile and mortar.

To start, locate the centerline of the flange, then measure and mark a point 12 in. from the bottom plate, plus the thickness of any finished wall you haven't installed yet. This measurement will typically total 12½ in. (assuming ½-in. drywall). That mark represents the center of the hole for the flange pipe that goes through the floor. If the hole's too small, the pipe won't fit. If it's too large, you won't have any wood to hold the screws **1**.

A CD is close to the proper diameter and makes an easy template. In fact, if you're using a 3-in. pipe flange (the most common), and the flange pipe bevels, the outside diameter where it meets the floor is about 4¾ in.—the diameter of a CD **2**. For flanges without a bevel, use the CD to make a reference circle, then center your flange inside that circle to mark the pipe diameter **3**. Drill a ³⁄₈-in. hole with a twist drill so its edge falls just

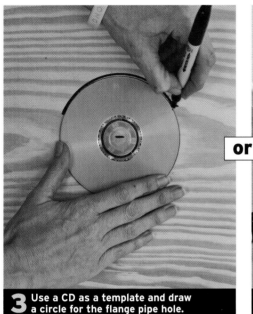

or

3 Use a CD as a template and draw a circle for the flange pipe hole.

If the circle is too big, center the pipe inside the CD mark and outline it.

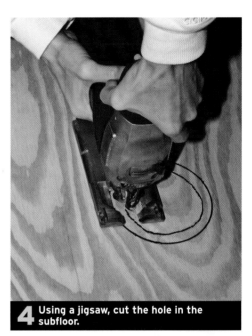

4 Using a jigsaw, cut the hole in the subfloor.

8 Keep the bowl hold-down bolts from falling over with a washer and nut (top).

Picking a toilet flange
Toilet flanges are like plungers, there are all kinds of them, but there's only one you need. It's all plastic (no rotating parts), it's open in the center (no knockouts), and has two small horseshoe cutouts on each side. Ignore the crescent-shaped slots for the bowl bolts; they can let the bolts spin when you try to turn the nuts.

There are myriad flanges to confuse you, yet you need only one—the one in the very center, which is called a 4-in. by 3-in. flange.

on the inside edge of the circle, then insert a rough-cut jigsaw blade and carefully cut the circle out, turning the saw constantly to keep the blade on the line **4**. Insert the flange to verify the proper hole diameter. Orient the flange so the horseshoe-shaped cutouts on each side are at dead center—each 12½ in. from the wall behind the toilet. Holding the flange in place, mark the location of the flange screws **5**. Don't use the arced cutouts

on the flange; they can allow the bowl hold-down bolts to spin when you try to remove them at some later date. Remove the flange and, using a drill bit about half the diameter of a #12 by 1½-inch flat-head wood screw, predrill the holes at your marks **6**. Predrilling makes driving the screws easier and keeps the wood from splitting. Line up the flange so the mounting holes match with the bowl hold-down slots parallel to the back wall, and drive

in the screws with a cordless drill **7**. Install any remaining fixtures and the finished wall and floor. Rather than sliding the bowl hold-down bolts into their slots and having them fall over as you try and drop the bowl on the flange, keep them vertical with a washer and nut **8**. Then mount the toilet.

➡ See "Replacing a Toilet," pp. 132-135.

KITCHEN APPLIANCES

ONE OF THE MORE CONVENIENT aspects of hooking up kitchen appliances is that unless you've embarked on a complete remodel, there's not much rough plumbing you have to do.

Most kitchens have only one of each appliance—a sink, a dishwasher, and a refrigerator—and the rough plumbing is already in place for those items. So the one drain line you need and the single pair of water-supply lines are already there. Adding new appliances like a dishwasher or an ice maker requires the addition of only flexible supply lines and perhaps a few adjustments in drain configurations.

The rest is easy, and you'll probably find the hardest part lies in restraining your impulse to get things done quickly. Connections have to be glued or snugged with a wrench, pipe measurements have to be exact, and cut lines right on the money. All this takes more patience than skill. Just a little discipline and time and you're done.

CONVENIENCES

WATER QUALITY

SINKS

DISPOSALS AND DISHWASHERS

INSTALLING AN ICE-MAKER LINE

An ice maker is one of those luxuries you just can't live without once you have it. You quickly get used to serving iced drinks whenever you want them, especially when you can enjoy both crushed ice or cubes. Once you have the ice maker, you need to hook it up. Out of the back of the fridge you'll see a fitting or small tube you will connect to an ice maker pipe, which in turn connects to a stop valve in a valve housing in the wall or through a hole in the floor to a valve in the basement water line. An ice-maker kit, on the other hand, comes with everything you need—pipe, box, and valve. It's the easiest installation method. It does, however, require you to cut in a T-fitting into a cold water line and take a small-diameter line to the valve in the box.

HOOK-UP OPTIONS

You have a number of ways to hook up your ice-maker water line. The most common are shown below.

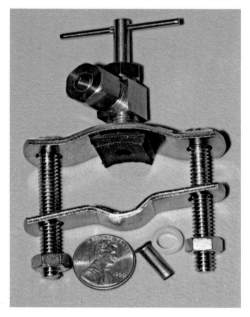

With a saddle valve, you can go straight from the water line to the fridge and not use a valve housing.

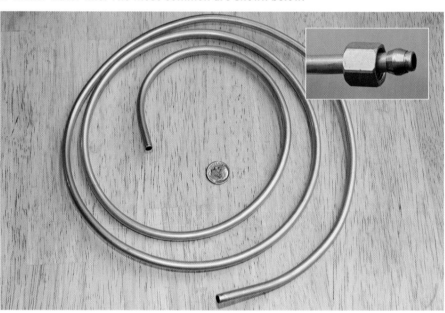

A copper pipe will need a nut and ferrule on both ends.

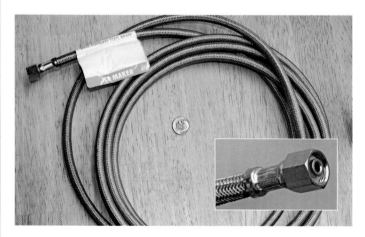

Braided ice-maker tube is the best line for ice makers, and comes with the connectors already attached.

Ice-maker kits contain all the materials you need to install a water line to the fridge.

➜ See "Installing T-Fittings," p. 52.

Typically the valve has a ½-in. threaded shank that takes ½-in. female fittings ❶, or you can sweat a copper extension directly to the inside of the shank ❷. Then mount the box, using 2x spacers called "scabs" to close up the stud opening to the width of the housing flanges ❸. Connect the pipe to the interface fitting you installed on the threaded shank. Now it's time to connect the tubing. Connect one end of the supply pipe to the valve ❹, and the other side to the fridge ❺. If the fridge has a cable, connect to it instead. You may need a coupling if one isn't on the fridge's tubing. Tighten all connections, turn the water on at the valve, and check for leaks ❻. You will have ice in a few hours, once all the air has been purged from the line.

1 Install an interface fitting to the house plumbing on the threaded shank of the valve.

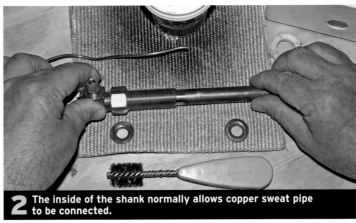

2 The inside of the shank normally allows copper sweat pipe to be connected.

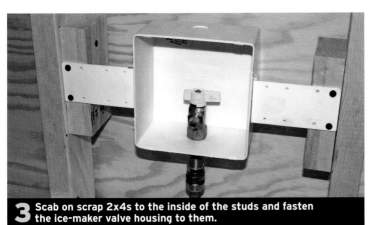

3 Scab on scrap 2x4s to the inside of the studs and fasten the ice-maker valve housing to them.

4 Connect one end of the supply pipe to the valve.

5 Connect the opposite end of supply pipe to the fridge.

6 Turn on all valves, and check for leaks.

INSTALLING A HOT WATER DISPENSER

1 Mount the water heater under sink (here a board is used as a pseudo wall).

2 Cut the hot and cold water hoses to length and attach them to the designated ports on the water heater.

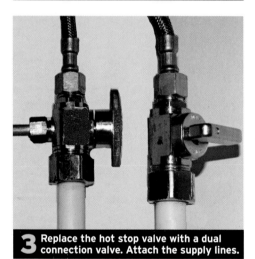

3 Replace the hot stop valve with a dual connection valve. Attach the supply lines.

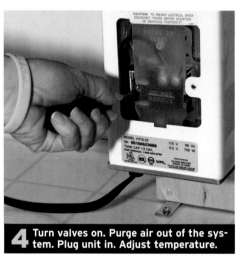

4 Turn valves on. Purge air out of the system. Plug unit in. Adjust temperature.

WARNING
Water from a hot water dispenser is scalding; do not touch it as it comes out of the faucet head—as it poses the risk of severe burns.

A hot water dispenser gives you instant hot water for hot drinks like coffee, tea, or hot chocolate, and you can use it to flush away dried food on dishes before you put them into the dishwasher. It does require, however, an unused hole on the sink rim, and if you don't have one, you can remove the sprayer or convert the kitchen faucet to a single-hole model. You can also drill or punch a new hole in the sink—but for most homeowners, that will prove impractical. You must also install a 120v grounded outlet under the sink to power the heating unit, if one isn't present already.

Start by turning off both the hot and cold water stop valves under the sink and removing the supply tubes. You will also need to shut off the hot water at the water heater. Place the heater unit under the sink close to where you will fasten it to the floor or side wall. If you can't get the unit off to the side, you can place the heater directly in the middle of the sink cabinet, using angle clamps and a board to create a pseudo wall **❶**. Following the manufacturer's instructions, cut the hoses to length and attach them to the heater and dispenser on the designated ports **❷**. Attach the water line to a T-cutoff

valve under the sink on the hot water side. Attach the faucet supply tubes back on the stop valves **❸**. Turn the stop valves and the hot water heater valve on and check for leaks. Adjust the thermostat to the desired temperature. Make sure the hot water spout is pointed to the sink center and that the main kitchen spout can swing over to the sink bowl. Turn the dispenser handle on to purge the air and plug the unit in **❹**. Failure to purge out all the air before you plug the unit in may result in a blown element. Wait about one hour to make your hot cocoa.

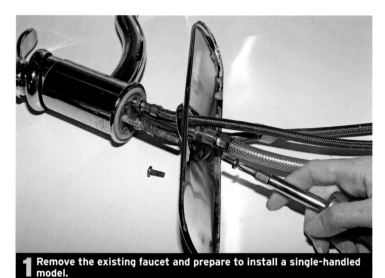

1 Remove the existing faucet and prepare to install a single-handled model.

Creating space with a single-handled faucet

Pull the existing faucet from the sink and remove the chrome escutcheon from its bottom ❶. Using a razor blade, clean off the area under the old faucet ❷. If a hole on the left or right side of center already exists, then the sink is ready for the new hot water dispenser faucet. If not, you will have to drill one, a project for which you may need to call in a pro. If you wind up with an extra hole, plug it with a chrome cap outfitted with spring "fingers" and made specifically for this purpose. Here, the existing faucet will use the center hole ❸, and the hot water dispenser will use an existing hole on the left ❹. Install the single faucet, and then the hot-water dispenser as described on the facing page.

➤ See "Installing a Single-Handled Bath Faucet," p. 145.

2 Clean any debris or sealant that was left from the old faucet.

3 Install the new faucet in the center hole.

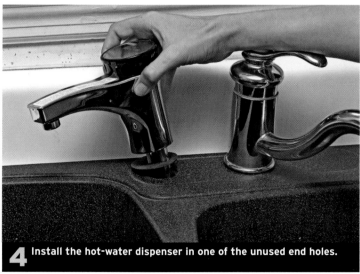

4 Install the hot-water dispenser in one of the unused end holes.

Make sure the hot-water spout does not interfere with the swing of the faucet spout.

INSTALLING A KITCHEN SINK SPRAYER

When you buy a kitchen faucet, you normally have a choice of whether you want a sprayer with it or not. Of course, even if you get a sprayer, you don't have to hook it up—you can plug off its outlet and the hole in the sink rim where it would install. But don't put off buying or installing a sprayer out of any fear of its installation. Typically, hooking a sprayer up is quite simple, but is best done before you install the main faucet. You start by laying a bead of silicone caulk around the sprayer hole to the right of the main faucet. Insert the sprayer housing into the hole ❶, and tighten it down from underneath with the manufacturer's nut ❷. Then push the end of the sprayer hose into the housing until the sprayer body seats in the housing ❸,❹. Pull the sprayer hose back up through the center faucet hole ❺. Wrap the male faucet threads with some turns of Teflon tape ❻, and attach the hose to the faucet—first with your fingers and then with a wrench ❼. Install the faucet, hook up the water lines, turn the water on, and check for leaks.

➡ See "Installing a Single-Handled Bath Faucet," p. 145.

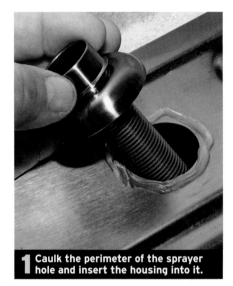

1 Caulk the perimeter of the sprayer hole and insert the housing into it.

2 Working from under the sink rim, secure the sprayer body with the nut.

3 Insert the sprayer hose into the housing.

4 Seat the sprayer body in its housing.

5 Pull the sprayer hose through the center hole of sink.

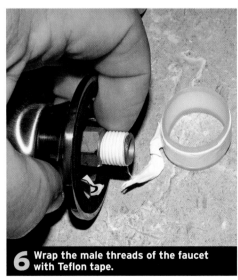

6 Wrap the male threads of the faucet with Teflon tape.

7 Attach the sprayer hose to the faucet, and secure it with a wrench.

INSTALLING A REVERSE-OSMOSIS FILTER

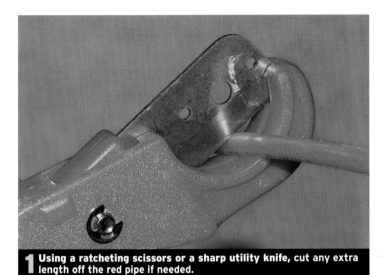

1 Using a ratcheting scissors or a sharp utility knife, cut any extra length off the red pipe if needed.

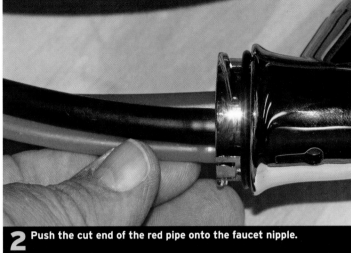

2 Push the cut end of the red pipe onto the faucet nipple.

3 Slide the faucet pipes and attachment screws into the hole in the sink rim.

If your sink rim doesn't have a hole for the water spout, **drill a hole** (typically 1¼ in.) through the rim and the countertop. Otherwise, use the available empty hole.

A reverse-osmosis (RO) system will filter your water to a semipure state (but not as pure as distilled water) using a set of three cartridges: a prefilter, a membrane filer, and a postfilter, all of which will need replacing as time goes by. One of the main requirements, therefore, is that you mount the cartridges in a location that leaves enough room under them to remove and replace them.

The system can be installed under a kitchen sink if there is room in the sink cabinet, on a block wall in a basement, or on any wooden wall. If mounted at a location other than under a sink, it will need a water supply and a drain (such as a clothes-washer drain or a floor drain). It will send around 35 gal. of water down the drain for 1 gal. or 2 gal. of semipure water.

If you need a new hole in the sink rim for the RO faucet (as opposed to using the sprayer hole), drill or punch a hole of the size recommended by the manufacturer (typically 1¼ in.). You may want to call in a pro for this operation or a plumber friend, because the chances of damaging the sink are high.

➔ See "Installing a Basement Reverse-Osmosis Filter," p. 213.

There is a small red hose that comes off the filter block—attach it to the faucet. If needed, it is possible to cut out any excess hose length from the middle **❶**,**❷**. Once the faucet is mounted, you will connect any cut ends back together with a ¼-in. OD push-on coupling (bought separately). Slip the faucet into the hole on the sink. Insert the pipes first and then slide the faucet in, taking care not to scratch the sink with the two attachment screws and wing nuts **❸**. >> >> >>

INSTALLING A REVERSE-OSMOSIS FILTER (CONTINUED)

4 Lift up the faucet mechanism slightly to expose the attachment screws, and tighten them.

5 Orient the RO faucet in the proper direction before tightening down the wing nuts from the underside of the sink.

Turn the top of the RO faucet 90 degrees and lift it slightly off to one side to expose the heads of the two tightening screws. Tighten them down **4**. Make sure the wing nuts under the sink do not pinch any pipes as they tighten **5**. If the RO unit uses a battery to power its electronics, install the battery, following the manufacturer's instructions **6**. Under the sink, connect the two cut ends of the red hose back together with a ¼-in. OD push-on coupling. Push the hose into the coupling until you feel it bottom out **7**.

To mount the set of cartridges, drive in two screws spaced to match the distance there is between the two hanging slots on the cartridge back. Set the tank adjacent to the cartridges, and connect the colored hoses to their respective mounts per the manufacturer's instructions **8**. The RO drain will consist of a slip-in piece with a black grommet and a common stand pipe. Insert the slip-in piece into the side of the stand pipe **9**, and fasten it with the supplied nut **10**. Insert the RO drain onto the kitchen drain immediately after it turns down to the P-trap. Insert the black drain into the black grommet on the RO drain. If needed, shorten the sink drain stand pipe. This will fit into the drain system on either an end drain or a center drain **11**.

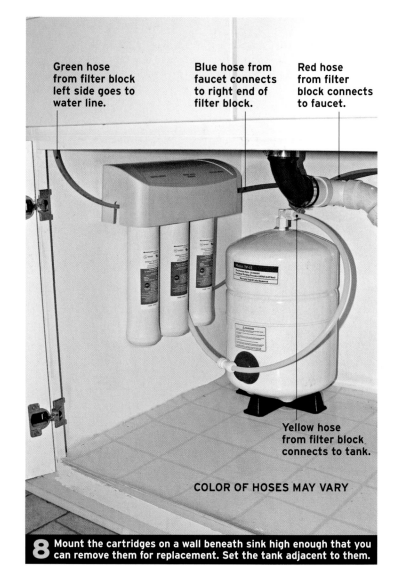

Green hose from filter block left side goes to water line.

Blue hose from faucet connects to right end of filter block.

Red hose from filter block connects to faucet.

Yellow hose from filter block connects to tank.

COLOR OF HOSES MAY VARY

8 Mount the cartridges on a wall beneath sink high enough that you can remove them for replacement. Set the tank adjacent to them.

6 Install the battery in its housing, following the manufacturer's instructions.

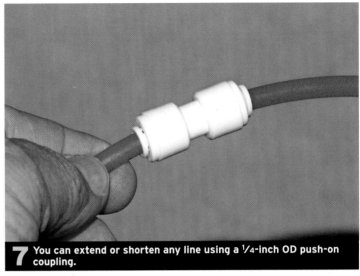

7 You can extend or shorten any line using a ¼-inch OD push-on coupling.

9 Insert the RO drain into the side of the standpipe.

10 Tighten the RO drain with its nut.

or

11 Install the RO stand pipe above the existing P-trap, and insert the free end of the RO drain line into the stand pipe fitting.

Insert the RO drain above the P-trap into a center drain assembly.

INSTALLING A KITCHEN SINK

1 Install the countertop on the base cabinet and set the sink upside down on the counter at its center line. Outline the perimeter of the sink on the countertop.

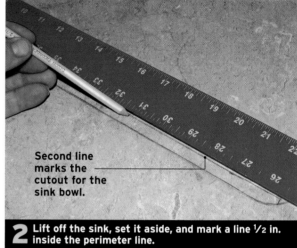

Second line marks the cutout for the sink bowl.

2 Lift off the sink, set it aside, and mark a line ½ in. inside the perimeter line.

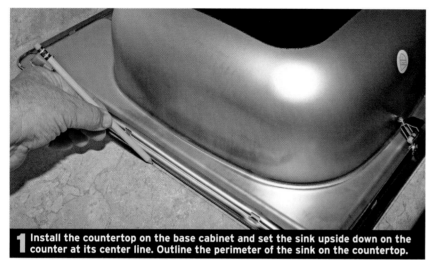

Edge of spade bit hole just to the inside of the cutout line.

3 Drill out the corners at the second line just to the inside edge of the line.

Cut out just to the inside of the second line.

4 Using a jigsaw with a fine-tooth blade, insert the saw in one hole and cut along the inside edge of the second line.

At first, installing a kitchen sink in a new countertop might seem a bit overwhelming. But when you get down to it, it involves nothing more than cutting a rectangular hole in a piece of wood. Often its location is centered under a kitchen window, and this automatically gives you a centerline for the sink. All you have to do is install the countertop, then place the sink (or template) upside down on that centerline, and outline its perimeter on the countertop surface **1**. If you are using the sink as your template, draw another line ½ in. inside the first line **2**. Cutting along the inside line will allow the bowls to fit snugly, with the rim resting on top of the counter. Before you cut the inside line, open the corners with a spade bit to provide a starter hole and turning room for the jigsaw blade **3**. Carefully cut the hole along the inside of the cut line, not on it **4**. When using a sabersaw or jigsaw on a finished surface, cut with a fine-tooth blade. That will reduce chipping on the surface to about ¹⁄₁₆ in., chips that the sink lip will easily hide. Caulk around the cut **5**, then set sink in the hole **6**. From underneath, adjust the clamps so they pull the sink lip down **7**.

5 Lay a bead of silicone caulk around the cut edge.

6 Set the sink in the hole and press it evenly into the caulk.

7 From underneath, tighten the clamps. Remove excess caulk from the lip edge.

Cutting from the bottom

Cutting the kitchen sink hole from the bottom puts the counter top in a better position to work on. And you don't have to worry about scratching the counter's surface. The disadvantage is that cutting before installing weakens the center, which can snap the counter, so have a helper on hand when you put the counter in place. Follow the same steps as you would cutting from the top as shown on the facing page, finding the centerline, outlining the perimeter, marking an inside line, then drilling out holes in the corners, staying just to the inside of the second line. Using a spade bit, drill the start holes just until the tip of the pit pierces the counter surface **❶**. Then flip the counter over and finish the holes from the top **❷**. This will reduce chipping out the finished surface. Then cut the inside line with a sabersaw. Flip the countertop over, fasten it to the cabinet base, drop in the sink, and install it and the faucets.

1 Mark lines and drill the corners just until the tip of the bit pierces the finished side.

2 Turn the counter over, and complete the hole from the finished surface.

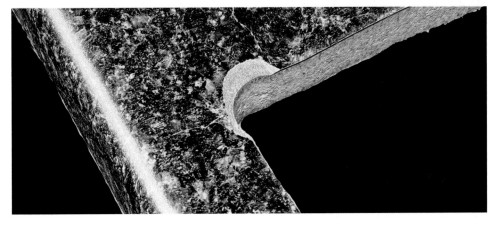

This countertop was drilled from the bottom **all the way through the finished surface. The blowout around the hole edge is obvious. Always complete the corner holes by drilling from the finished surface of the countertop.**

INSTALLING A KITCHEN SINK DRAIN

1 Insert a flat-head gasket into the upper fitting of the vertical T.

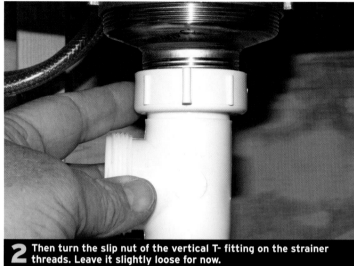

2 Then turn the slip nut of the vertical T- fitting on the strainer threads. Leave it slightly loose for now.

The drain system under the kitchen sink is typically tubular thin-wall pipe a little more than 1 in. in diameter. It is assembled with plastic compression fittings. The two most common configurations locate the P-trap on the end or in the center. The P-trap empties into the drainpipe entering the wall or floor. Both end and center drain fittings come threaded for either a direct connection to the strainer threads, or a slip-on connection to a short stand pipe between the fitting and the strainer. Always avoid the stand pipe configuration, it has a tendency to leak. And regardless of which design suits your installation, always use fittings that thread directly on the strainer. If you have a choice between the two drain layouts, the end drain is the better of the two because it is less prone to clogs and leaks.

An end drain will have two pipes, one vertical and one horizontal. The bowl you want the P-trap on will get the vertical pipe, the other bowl gets the horizontal pipe. Start by inserting one of the special gaskets with a flat on top into the vertical T drain-to-strainer connection ❶. Fasten the vertical T loosely to the strainer ❷. Fit the horizontal pipe at its elbow with a flat-head gasket and at the other end with a slip nut and tapered gasket, dry fit the parts, cut the horizontal pipe to length if necessary, and install the horizontal pipe ❸– ❻. Once assembled, retighten all the wing nuts and slip a P-trap on the bottom of the drain ❼. Install a PVC trap adapter or a flexible trap adapter as a transition between the sink drainpipe and the main drain line ❽. The flexible model allows some give in the connection angle; the PVC adapter has to be installed at an angle that's dead on the money.

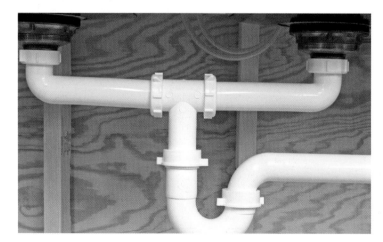

Kitchen drains can be configured **with a P-trap centered between the two bowls of the sink.**

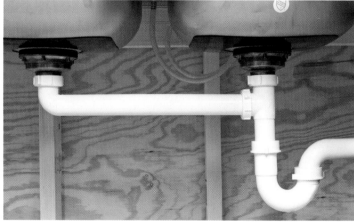

You can also set up a kitchen sink drain **with a P-trap mounted directly under the strainer of one bowl. This configuration is easier to install.**

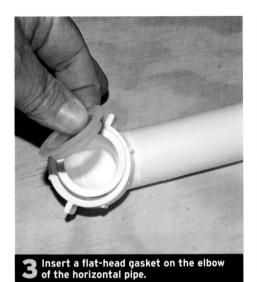

3 Insert a flat-head gasket on the elbow of the horizontal pipe.

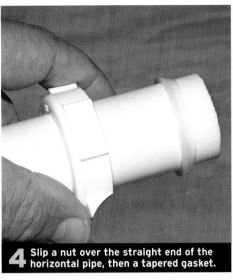

4 Slip a nut over the straight end of the horizontal pipe, then a tapered gasket.

5 Insert the end of the horizontal pipe into the T-fitting.

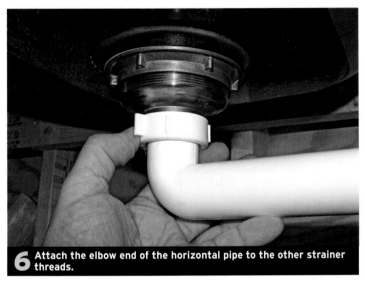

6 Attach the elbow end of the horizontal pipe to the other strainer threads.

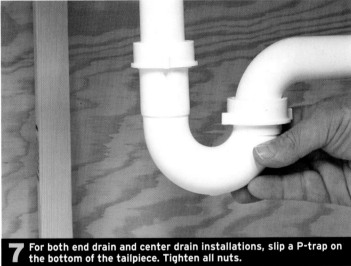

7 For both end drain and center drain installations, slip a P-trap on the bottom of the tailpiece. Tighten all nuts.

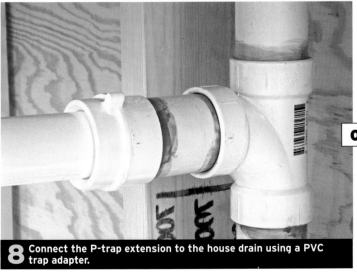

8 Connect the P-trap extension to the house drain using a PVC trap adapter.

or

Connect the P-trap extension pipe into the house drain with a flexible trap adapter, best if the connections are slightly off-angle.

INSTALLING KITCHEN SINK STRAINERS

1 Apply a bead of silicone caulk to the drain flange in the sink.

2 Apply another bead of caulk to the strainer lip and push it into the caulk.

3 Push on the gasket and washer on the strainer from beneath the sink.

4 Insert the open jaws of needle-nose pliers in the strainer basket holes.

5 Thread on the strainer nut, hand tight. Tighten the nut with a spanner wrench.

6 Some strainers use lock bolts for a secure attachment.

You can install kitchen sink strainers either before or after you install the sink. Installing them first will prove somewhat easier because it requires less stooping. Start with a bead of silicone caulk around the sink's strainer hole and underneath the strainer's lip **1**. Push the strainer into the caulk (remove the excess later) **2**. From underneath, insert the rubber gasket and cardboard washer, if any **3**. From on top, insert the open jaws of needle-nose pliers through the holes **4**. From underneath, tighten the nut with a spanner wrench with one hand while keeping the strainer from turning with the pliers **5**. If the strainer nut is designed for it, lock it with the supplied bolts **6**.

INSTALLING A CUSTOM KITCHEN SINK DRAIN

The problem with kitchen sink drains is that they are designed to leak. One bump on any of the slip joints, and the drips begin. And, of course, sink drains are also made to obstruct everything stored under the sink.

There is a better way. You can take the drains straight back to the wall with flexible elbows and keep it all back there—trap and everything. Supported against the wall, a hammer blow won't make it leak. Start by removing all the old drainpipes and throwing them away. Attach a flexible elbow to each strainer, and point these straight toward the back wall ❶. Dry-fit a PVC elbow to a 1½-in. PVC pipe cut just long enough to fit in the flexible elbow and still leave the elbow

against the wall. Point the PVC elbows at each other ❷. Assemble a double-Y fitting with street 45 fittings on each side. Slip a glue-in bushing with ½-in. female threads in the top for a dishwasher drain (a ½-in. nipple will screw into the fitting, and the dishwasher drain will attach to that) ❸. Slip the double-Y assembly directly against the back wall between the two PVC elbows, connecting them with two short PVC pipes ❹. Out of the bottom of the double-Y, fit a small length of 1½-in. PVC as a connection for a schedule 40 P-trap ❺. Once all the fittings dry-fit together nicely, glue them and then strap them against the wall. Connect the P-trap into the house drain.

1 Attach flexible elbows to each strainer and point them toward the back wall.

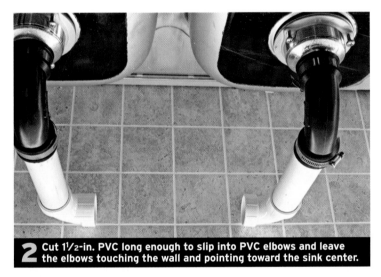

2 Cut 1½-in. PVC long enough to slip into PVC elbows and leave the elbows touching the wall and pointing toward the sink center.

Glued bushing for dishwasher drain

Street 45

Double-Y fitting

3 Assemble a double-Y fitting with street 45s and a ½-in. nipple for the dishwasher drain.

Short lengths of PVC pipe inside hubs connect elbows and street 45s.

4 Attach the double-Y fitting to the two elbows.

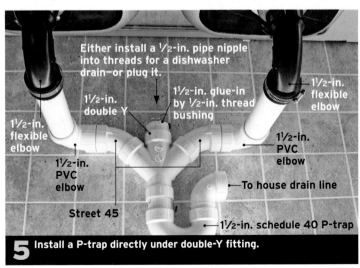

Either install a ½-in. pipe nipple into threads for a dishwasher drain—or plug it.

1½-in. double Y

1½-in. glue-in by ½-in. thread bushing

1½-in. flexible elbow

1½-in. flexible elbow

1½-in. PVC elbow

1½-in. PVC elbow

To house drain line

Street 45

1½-in. schedule 40 P-trap

5 Install a P-trap directly under double-Y fitting.

INSTALLING A GARBAGE DISPOSAL

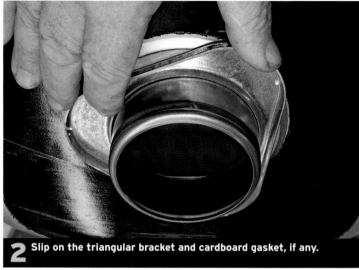

1 Remove the strainer if present, clean the opening, and caulk the flange. Insert the disposal flange into the strainer hole.

2 Slip on the triangular bracket and cardboard gasket, if any.

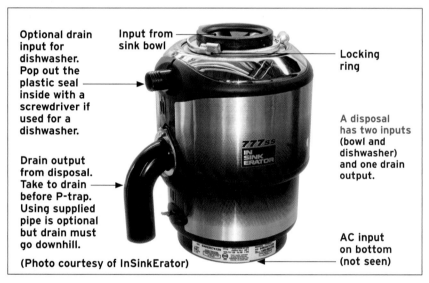

Optional drain input for dishwasher. Pop out the plastic seal inside with a screwdriver if used for a dishwasher.

Input from sink bowl

Locking ring

Drain output from disposal. Take to drain before P-trap. Using supplied pipe is optional but drain must go downhill.

A disposal has two inputs (bowl and dishwasher) and one drain output.

AC input on bottom (not seen)

(Photo courtesy of InSinkErator)

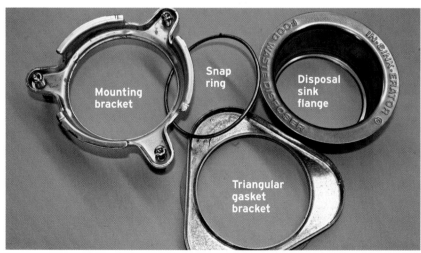

Mounting bracket

Snap ring

Disposal sink flange

Triangular gasket bracket

The disposal **mounting system.**

You can install a disposal under either strainer in a kitchen sink, but the common location is on the strainer opposite the drainpipe entry. Disposals will need a dedicated switched 120v outlet under the sink—not a line tapped into the kitchen counter receptacles. It is best if the disposal has its own branch circuit, but if another circuit is readily available (not the counter circuit) you could tap into it. Before you start installation, you need to plan the switch location. You can install a switch in the wall above the counter, an option that means cutting into the walls. An easier alternative is to get a disposal with the switch built into the lid. Even better is a disposal with an air switch. You mount the switch on the sink (like a push button) and it does not require running new electric lines.

Installing the disposal unit

Locate all the parts for the disposal locking mechanism that fastens to the strainer hole. Then remove the strainer if it's present on an existing sink, and clean the lip where the strainer was mounted. Caulk the strainer-hole flange and slip in the disposal's sink flange, orienting any lettering or logo as you wish **❶**. Holding the flange in place from above, slip on the triangular bracket and gasket **❷**. The round mounting bracket with three bolts goes on next, but you have to

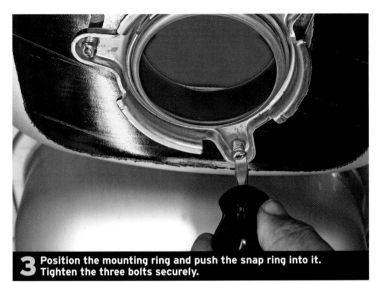

3 Position the mounting ring and push the snap ring into it. Tighten the three bolts securely.

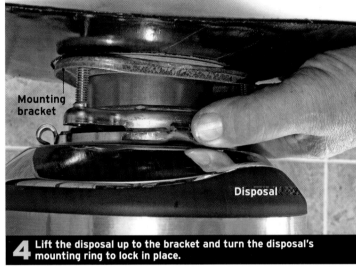

Mounting bracket

Disposal

4 Lift the disposal up to the bracket and turn the disposal's mounting ring to lock in place.

back off the three bolts (don't remove them completely) before you slip it on. Fit this unit on and slip on the snap ring clamp. Tighten the three threaded bolts with a screwdriver **3**. Lift the disposal into place under the bowl and twist the locking ring to secure it **4**. Make sure all three tabs engage the locking ring. All three sides have to slide up a flat metal arm.

>> >> >>

A disposal does not adversely affect a septic tank or drain lines. However, for those who need an extra measure of efficiency, InSinkErator makes a disposal that adds a biodegradable liquid to the ground food. (Photo courtesy InSinkErator)

Locking the disposal To lock the disposal in place, the locking ring must slide up a fixed metal ramp. You can slide the locking ring part-way up the ramp by hand, but you need the antijamb wrench to get it seated all the way. The disposal is designed with three slots for the wrench 120 degrees apart. Typically, if you insert the wrench in one slot and pull it, the entire ring will slide the last half inch to the locked position.

INSTALLING A GARBAGE DISPOSAL (CONTINUED)

Installing the drain

For a two-bowl sink, don't use the black curved pipe that came with the disposal. Instead, transfer the gasket from that pipe to a straight thin-wall pipe with one flat rounded end (buy separately) **❺**. Cut the pipes to fit between the disposal and the second strainer, keeping in mind the amount that has to be inserted into the fittings **❻**. Connect the long tube to the disposal **❼**, and the short tube to the strainer **❽**.

Wiring the disposal

The average disposal can be wired with either 12-gauge or 14-gauge (grounded) cable, unless it has a high horsepower rating and the directions specifically require 12-gauge cable. Make sure the breaker matches the cable: 14-gauge cable needs a 15-amp breaker; 12-gauge cable needs 20-amp protection. The cable can come up through the floor or out of the wall. A typical switched wiring scheme is illustrated on the facing page.

Connect the wires of similar color—black to black, white to white, and ground wire to the green wire or disposal body. You do not need to use flex metal armor unless your local codes require it. Splice the switch into the black or hot wire of the incoming cable, never switch the white or neutral conductor.

5 Transfer the rubber gasket from the curved disposal pipe to the straight.

6 Cut the drain tubes to proper lengths so they fit between the disposal and the T.

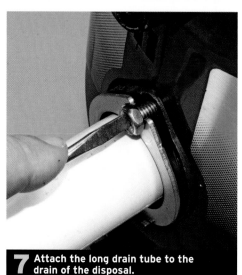

7 Attach the long drain tube to the drain of the disposal.

8 Attach the short tube to the strainer threads and both tubes to an end drain.

To unjam a stuck disposal, insert the hex wrench into the recess in the bottom center of the disposal and turn it to break a jam free.

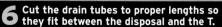

The drain assembly: A long horizontal arm fits into the side of the disposal, and a short vertical pipe attaches to the strainer. Both of these pipes will then slip into a T-fitting with a P-trap on the bottom.

WIRING A DISPOSAL

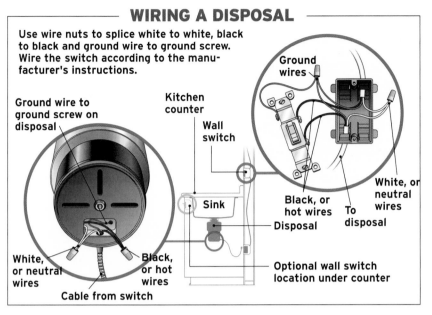

Use wire nuts to splice white to white, black to black and ground wire to ground screw. Wire the switch according to the manufacturer's instructions.

Ground wire to ground screw on disposal

Kitchen counter

Wall switch

Ground wires

White, or neutral wires

Black, or hot wires

To disposal

Sink

Disposal

White, or neutral wires

Black, or hot wires

Cable from switch

Optional wall switch location under counter

A single-bowl sink will use a curved black drain (which comes with the disposal) that goes right into a P-trap.

Remote push-button disposal switch

Forget hard wiring within a wall. Forget cutting into a wall to mount an outlet box. You can remotely control a disposal with an attractive push button mounted to the sink or countertop. When you push the button, it sends a pulse of air through a flexible tube to the control box mounted under the sink. The pulse of air closes a switch, which allows power to go to the disposal. You will, however, have to mount an electrical receptacle under the sink if one is not already there. Hidden under the sink, this surfaced-mounted electrical box can be installed with far less effort than a flush box on the kitchen wall.

An air-powered switching unit saves you time and trouble in installing a switching mechanism for your disposal.

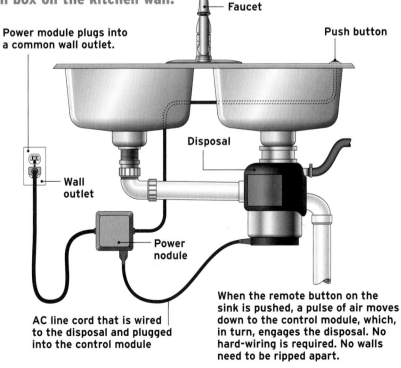

Faucet

Push button

Power module plugs into a common wall outlet.

Wall outlet

Disposal

Power nodule

AC line cord that is wired to the disposal and plugged into the control module

When the remote button on the sink is pushed, a pulse of air moves down to the control module, which, in turn, engages the disposal. No hard-wiring is required. No walls need to be ripped apart.

INSTALLING A DISHWASHER

For anyone who faces dish washing chores on a regular basis, a dishwasher is your best friend. Most dishwashers are installed to the immediate right of the sink because most people are right handed. Dishwashers are designed to fit under a kitchen countertop. Most models need a cabinet enclosure 24$\frac{1}{4}$ in. to 24$\frac{1}{2}$ in. wide.

Installing the water lines

In the side of the dishwasher cabinet nearest the sink and in the back just below the countertop, drill a 1$\frac{1}{2}$-in. hole for the drain line. Then drill a 1-in. hole in the bottom of the same side for the water line ❶. Very gently, lay the dishwasher on its back and remove the fasteners (Phillipshead screws or $\frac{1}{4}$-in. hex head screws) that hold the kickplate (the plate immediately below the door) ❷. Set the kickplate aside and do not bend or scratch it. This exposes the area in which you'll find connections for the water and electrical lines; the water line solenoid is on the left and electrical connections are on the right. Make sure the drain hose is already attached to the dishwasher pump; if yours is not, attach it with the clamps that came with the washer for this purpose ❸.

Look for the dishwasher solenoid at bottom left. You'll recognize the $\frac{3}{8}$-in.-female thread attached to it. That's where the water line fitting goes. Hand tighten the water line fitting (a $\frac{3}{8}$-in. compression angle) to the

interface on the solenoid. Then, using an adjustable wrench, tighten the fitting so it faces toward the back of the dishwasher and slightly to the left. Attach flexible dishwasher water line (a 4-ft. to 5-ft. length of $\frac{3}{8}$-in. OD supply tube) to the water line fitting ❹. Do not use a rigid water line; it will not allow you to pull the dishwasher out for maintenance. First hand tighten the water line, then snug it with a wrench.

Installing the electrical lines

Remove the cover of the electrical box located near the front right bottom of the dishwasher. Attach an NM bushing in the cable hole on the electrical box. Bring the ends of the cable (12-gauge NMB cable with ground) through the bushing, and splice the wires to those already in the dishwasher. Use orange wire nuts (they're the right size for this wire) to connect the wires of the same color—black to black, white to white, and ground to a metal ground screw or green wire ❺.

Finishing the installation

Return the dishwasher to its upright position, and start to feed the ends of the electrical cable, water line, and drain line through the holes you cut in the cabinet. Then push the dishwasher into the cabinet while a helper pulls the lines into the sink cabinet next door. You can let electrical cable and any extra water line loop up behind the dishwasher.

>> >> >>

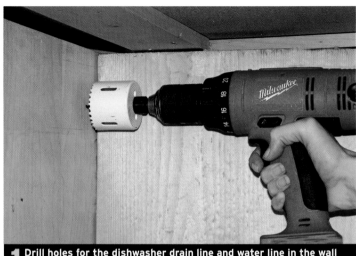

1 Drill holes for the dishwasher drain line and water line in the wall adjacent to the sink.

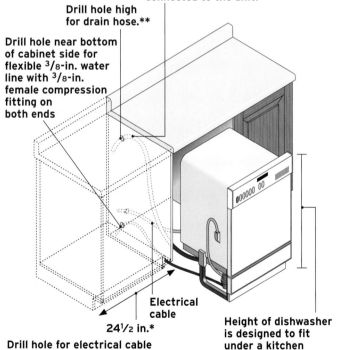

TYPICAL DISHWASHER INSTALLATION

Drain line comes with dishwasher and may be already connected to the unit.

Drill hole high for drain hose.**

Drill hole near bottom of cabinet side for flexible $\frac{3}{8}$-in. water line with $\frac{3}{8}$-in. female compression fitting on both ends

Electrical cable

24$\frac{1}{2}$ in.*

Drill hole for electrical cable behind dishwasher or bring cable in from kitchen sink cubicle. Use 12-gauge NMB cable with ground on a dedicated 20-amp circuit.

Height of dishwasher is designed to fit under a kitchen counter (it is the same height as the base cabinetry).

*Or, 18$\frac{1}{4}$ in. for space-saver dishwasher.
** Some codes may require an air break filling.

2 Lay the dishwasher on its back, and use a Phillips screwdriver or a nut driver to remove the screws holding the kickplate.

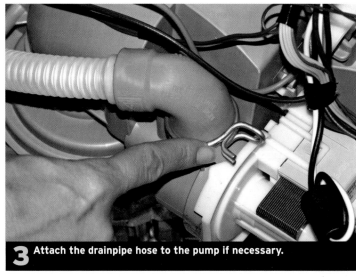

3 Attach the drainpipe hose to the pump if necessary.

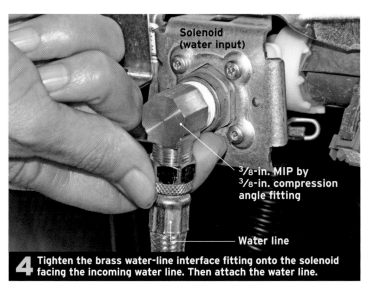

Solenoid (water input)

³/₈-in. MIP by ³/₈-in. compression angle fitting

Water line

4 Tighten the brass water-line interface fitting onto the solenoid facing the incoming water line. Then attach the water line.

5 Bring cable into the wiring box and connect the wires color by color to the dishwasher wires.

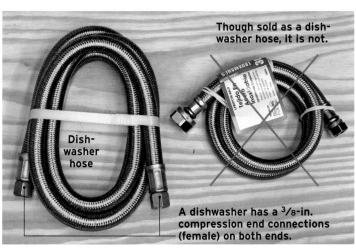

Though sold as a dishwasher hose, it is not.

Dishwasher hose

A dishwasher has a ³/₈-in. compression end connections (female) on both ends.

When purchasing a dishwasher water line, get flexible line with ³/₈-in. compression fittings on both ends.

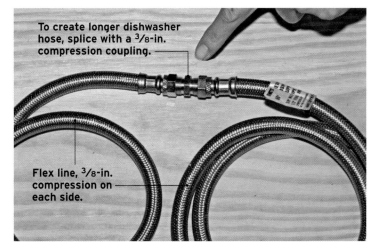

To create longer dishwasher hose, splice with a ³/₈-in. compression coupling.

Flex line, ³/₈-in. compression on each side.

If you need an extra-long water line, splice two flex lines together with a coupling.

INSTALLING A DISHWASHER (CONTINUED)

6 Slide the dishwasher into the cabinet and check for level.

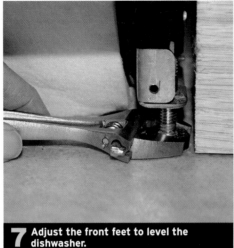

7 Adjust the front feet to level the dishwasher.

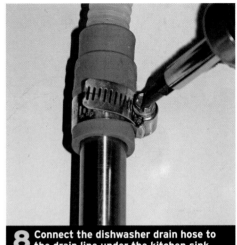

8 Connect the dishwasher drain hose to the drain line under the kitchen sink.

With the unit pushed completely in its cabinet, set a level on top of the front edge of the dishwasher **6**. Using a small open-end wrench, adjust the appropriate front leg(s) as necessary to bring the appliance level **7**. Only the front two feet should be adjusted.

Attach the dishwasher drain hose to the undersink drain—its end will slip on a ½-in. nipple of any pipe material **8**. Each drain installation, of course, will differ in how the hose will attach to the sink drain, because there are a variety of possible drain fittings. In our installation, we've used a ½-in. nipple screwed into a threaded PVC bushing and a double-Y fitting.

→ See "Installing a Custom Kitchen Sink Drain," p. 181.

Do not add a trap. A dishwasher uses the sink trap. Cut a T-fitting into the hot water line to the sink, add a stop valve, and attach the flexible water line fitting **9**. Bring the electrical cable to an electrical junction box on a dedicated 20-amp circuit. Then turn the water on and as the dishwasher fills, check for leaks **10**. If there are no leaks, secure the top of the dishwasher cabinet to the underside of the counter by driving the screws provided through the holes in the tabs on top of the dishwasher's body **11**.

WHAT CAN GO WRONG

If the dishwasher is not level, there will be an uneven level of water inside it, which may affect its cleaning action and the operation of the float switches.

DRAIN LINE FITTINGS

There are a variety of fittings available to tap into the thin-wall drain lines under the kitchen sink.

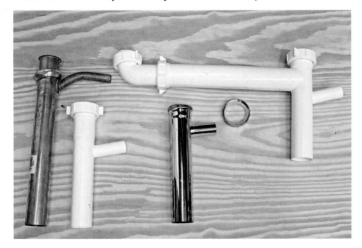

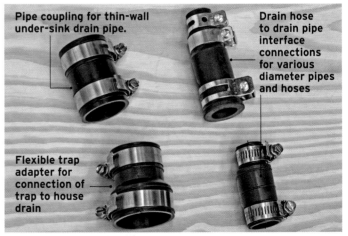

Pipe coupling for thin-wall under-sink drain pipe.

Drain hose to drain pipe interface connections for various diameter pipes and hoses

Flexible trap adapter for connection of trap to house drain

9 Install a separate stop valve in the hot water line or this dual-outlet valve.

To faucet

To dishwasher

Hot water line

10 Check for leaks under the washer, especially at the water line connection.

11 Screw the anti-tip bracket flanges into the bottom of the cabinet.

DRAINING A DISHWASHER

There are many ways to drain a dishwasher. If you are lucky, there will be a drain access in the existing drain. Or, if you have a disposal, the side drain is always there, but you have to use a screwdriver to break out the plastic knockout. If none of these is available, you will have to take apart the exiting drain and add a tail piece with access or replace the existing drain with an integral drain for the dishwasher.

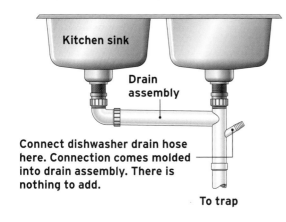

Kitchen sink

Drain assembly

Connect dishwasher drain hose here. Connection comes molded into drain assembly. There is nothing to add.

To trap

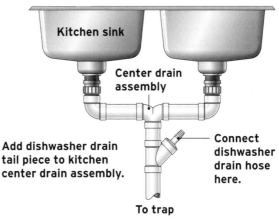

Kitchen sink

Center drain assembly

Add dishwasher drain tail piece to kitchen center drain assembly.

Connect dishwasher drain hose here.

To trap

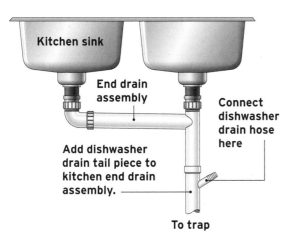

Kitchen sink

End drain assembly

Connect dishwasher drain hose here

Add dishwasher drain tail piece to kitchen end drain assembly.

To trap

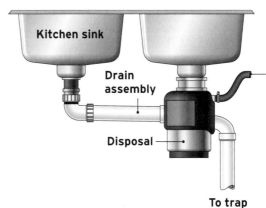

Kitchen sink

Drain assembly

Connect dishwasher drain hose here (must remove knockout on disposal). Connection comes molded into disposal.

Disposal

To trap

BASEMENTS AND UTILITY ROOMS

IT IS HERE IN BASEMENTS AND utility rooms that you'll find all the "wet" appliances, such as water heaters, washers, water softeners, and filters. These appliances need maintenance from time to time, and their basement or "utility" location means that if things get a little wet in the process, no harm has come to structural elements of a house.

Plumbing in these areas can be easier than in other rooms, too. Usually you don't have to work around existing walls to run new lines, and if new work or a maintenance job takes a little time, it's not going to interfere with an "important" room, such as a main-floor bathroom. Besides, these utility areas provide the perfect spot for upgrades, such as softeners and filters.

CONNECTING A WATER HEATER

Heating water has evolved from pouring stove-heated water into a large tub to modern automatic temperature-controlled appliances. Today, we have hot water at the turn of a handle, but every component on a water heater can break down and interrupt your service. With today's labor rates, you can exceed the cost of a water heater on a single service call. Finding out how a water heater works and how to repair it can take a little time but will prove to be a good investment in the long run.

How a water heater works

Although water heaters in some areas (particularly the northeast United States) are oil fired, most use electric current or gas (either natural gas or propane) to heat the water. But heating water is not the end of its job. The unit also has to store the hot water until you use it. So in addition to the heating system, the tank must be insulated.

Water enters a heater from the cold-water supply in your house. The inlet pipe is called a dip tube and allows water to fill the tank. In an electric heater, two heating elements (one in the upper third of the tank and one at the bottom), powered by your home 240v system, heat the water. In a gas unit, a gas-fired burner heats the water.

Because your domestic water is under pressure, hot water leaves the tank through the hot water supply pipes, flowing out at the faucet when you open it. At the same time, cold water enters the tank to keep its level constant. Both systems employ thermostats that sense the temperature of the water and call for heat when the temperature drops. Both systems also feature a temperature and pressure (T&P) relief valve, which opens if either the water temperature or the pressure exceeds a safe limit. The valve is connected to a pipe that runs down the outside of the tank, ending about 6 in. from the floor.

Because corrosion is the chief enemy of a water heater tank, a magnesium anode rod is inserted in the water. This rod is designed to corrode before the tank, extending the life of the unit. However, once the anode is completely used up, the water will go to work on the tank, and once a tank springs a leak, it's pretty much a goner. Check the anode every five years and replace it if necessary.

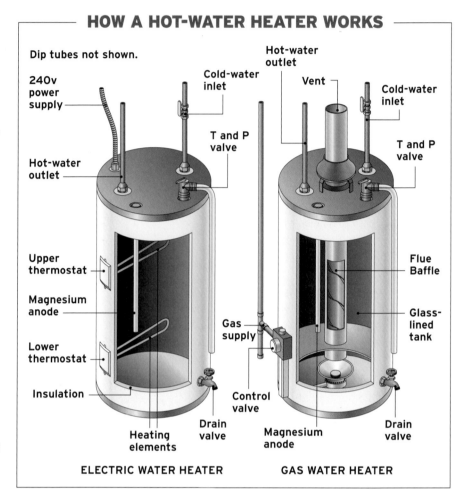

HOW A HOT-WATER HEATER WORKS

Dip tubes not shown.

240v power supply

Hot-water outlet

Cold-water inlet

Hot-water outlet

Vent

Cold-water inlet

T and P valve

Upper thermostat

Magnesium anode

Lower thermostat

Insulation

Heating elements

Drain valve

Control valve

Gas supply

T and P valve

Flue Baffle

Glass-lined tank

Magnesium anode

Drain valve

ELECTRIC WATER HEATER

GAS WATER HEATER

Sizing a heater As a rule of thumb, a family of four will need a 50-gal. electric or 40-gal. gas water heater. A smaller unit may save you money, but it can also make the sale of your house less desirable, since prospective buyers won't be satisfied with a less-than-standard water heater. Buy the most efficient water heater you can afford (it will have the thickest insulation). If you plan to live in your house forever, consider a lifetime warranty.

WHAT CAN GO WRONG

A 4,500w heater requires 10-gauge (with ground) cable. A 3,500w heater can get by with 12-gauge. Never run a 4,500w water heater on 12-gauge wire.

Venting

Electric water heaters do not require venting, but all gas water heaters must be vented to the outdoors to remove the by-products of combustion. The most common type of venting, atmosphere venting, sends the gasses vertically through the roof. Because the gases are warmer than the surrounding air, they rise naturally. Direct-vent or horizontally vented water heaters are designed for installation where vertical flues are not practical or too expensive to install. These vents go through an exterior wall. Power-vented gas water heaters use an electric fan or blower to send combustion gases to the outdoors.

An electric water heater can be installed almost anywhere, and a novice can do it. Not so with gas. Only a pro should install and maintain gas heaters because they require gas lines (which can explode), flues (which if improperly installed, can backdraft and kill you in your sleep), vent pipes, air resupply, condensation, makeup air, and enough codes to fill their own book. In addition, modern gas heaters have a sealed chamber, which prevents maintenance by homeowners.　　　　　>> >> >>

GAS-FIRED TANKLESS HEATER

A tankless water heater is designed to use electricity or (more commonly) gas only when there's a demand for hot water. This can be an economical choice, but only if hot water isn't needed from multiple fixtures at the same time.

Specific hookup and safety procedures will vary from unit to unit. Be sure to follow your manufacturer's instructions, including need for makeup air and type of flue. If you want to operate more than one faucet at one time, look for a unit that can heat at least 7gm. Super cold areas may require insulated dampers to keep cold air from freezing pipes. Hard water may adversely affect the pipes.

Need more hot water?

Whirlpool tubs, extra-long showers, or a large number of back-to-back evening showers can place a huge demand on a hot water system, which might not be able to keep up. You can always install one large water heater, but such heaters may be expensive, and you'll be heating a large amount of water at times when you won't be using it. That will send your utility bill sky high. The simple solution is to use two water heaters. The second heater (always electric) is controlled by a cutoff switch. The unit is left off most of the day; 1 hour before the extra water is needed, the switch is turned on, providing double the amount of available hot water. Once the need is over, the cutoff switch is turned back off.

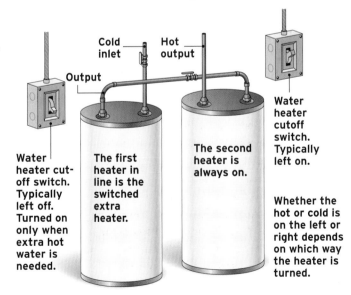

Cold inlet

Hot output

Output

Water heater cutoff switch. Typically left off. Turned on only when extra hot water is needed.

The first heater in line is the switched extra heater.

The second heater is always on.

Water heater cutoff switch. Typically left on.

Whether the hot or cold is on the left or right depends on which way the heater is turned.

An additional water heater with a separate electric switch will provide additional hot water when you need it. The additional heater is more efficient and less costly than a single large unit.

CONNECTING A WATER HEATER (CONTINUED)

Making the connections

You can connect your water heater tank to the supply system with straight pipe, such as copper, but using flex pipe offers some advantages.

First, it allows location flexibility. That is, when the roughed-in water pipes are not placed exactly above the water heater location, flex pipes will allow you to make such out-of-line connections. Next, female flex lines provide unions at each end, making installation and repairs quicker. In earthquake-prone areas, codes may require them, and they keep plastic pipe away from the heater connections, as required by code.

You'll find two types of flexible pipe: corrugated copper and stainless braided. Both work fine. However, the copper pipe will kink easily, and active water will cause it to pinhole. Though end terminators may vary, it's easiest to use female. Attach either end to the nipples coming out of the heater. Tighten the nut on the fitting with an adjustable wrench ❶. The opposite end of one pipe should connect to the heater valve on the cold side and the end of the other, directly to the house water line on the hot side ❷. Using closed-cell foam insulation, insulate the hot water pipe at least 3 ft. after it leaves the heater. You don't need to insulate the cold water pipe, since you want that water to warm to ambient temperature before it enters the heater.

➡ See "Insulating Water Lines," p. 228.

⚠ WARNING

If a gas heater is placed in a garage, be sure it is raised off the floor by around 18 in. or more to prevent gasoline vapor from going into the heater and exploding.

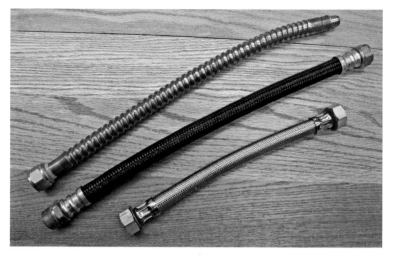

Flex pipes can be braided **or corrugated copper pipe with various types of ends. The latter can kink easily and will develop pinholes in aggressive water.**

1 Attach one end of the flex pipe to the nipple on the water heater.

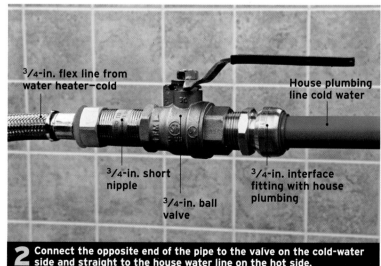

³/₄-in. flex line from water heater–cold

House plumbing line cold water

³/₄-in. short nipple

³/₄-in. ball valve

³/₄-in. interface fitting with house plumbing

2 Connect the opposite end of the pipe to the valve on the cold-water side and straight to the house water line on the hot side.

Installing flex lines on plastic insulating nipples

Some water heaters use plastic inserts to insulate the pipe nipple from the inside. These inserts, however, have a rounded edge, which prohibits the flex-line internal gasket from sealing properly.

In these cases, use a ³/₄-in. (brass or galvanized) metal coupling and nipple to create the alternate connection (the pipe threads will seal the union) ❶,❷,❸. The female end of a flex line will create a union on the nipple allowing easy removal of the tank for future maintenance.

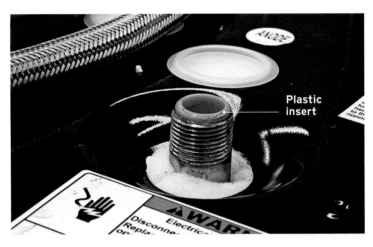

Plastic insert

A plastic insert inside a water heater nipple (used by the manufacturer to insulate the nipple) may create a leaky connection with flex pipe.

1 Tape the threads first and screw a galvanized coupling on the nipple.

2 Tape the threads of a short galvanized nipple and tighten the nipple into the coupling.

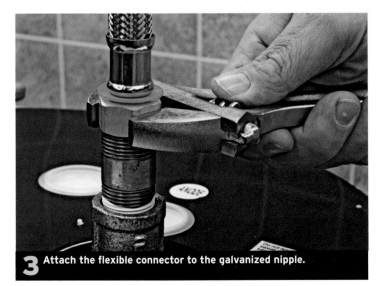

3 Attach the flexible connector to the galvanized nipple.

WARNING

If the water heater is placed on a finished floor, such as a wood floor on the second story, a leak can produce thousands of dollars of damage. It would be prudent to place the water heater in a water heater pan and run the drain of the pan to a floor drain or outside.

INSULATING A WATER HEATER

A water heater will start to lose heat the second the heat turns off. Tank insulation slows the heat loss, gives you more hot water to use, and saves you money. Although it's best to buy a heavily insulated new heater, a thinly insulated model will benefit from a jacket.

A water heater loses most of its heat at the floor, which acts like a heat sink. To break this contact, elevate a new heater with a couple 2x4s or concrete blocks ❶. If you want to raise an existing heater, you'll have to cut the pipes and install couplings.

To install the jacket, start at the T&P valve or element cover, with the top corner of the blanket even with the top of the heater and tape one edge of the jacket to the heater on top and bottom using duct tape ❷. Cut an opening to expose the element and thermostat covers ❸. Then bring the other edge of the jacket around the tank, cut an opening for the T&P valve, and tape this edge of the jacket to the first edge ❹,❺. Don't pull the jacket super tight. Fiberglass needs trapped air for insulation. Trim any excess jacket along the bottom of the tank. Then cut slits in the top piece for the water lines and the T&P valve (if it's on top) ❻. Tape the edges of the top piece to the side piece, cutting it in a circle or leaving it square ❼. Tape across the slit on top to close it.

1 Raise the heater to keep the floor from acting as a heat sink.

Water heater insulating jackets are made of fiberglass with a backing and come in two pieces: a blanket for wrapping the water heater and a rectangular piece for the top.

2 Tape one edge of the insulation jacket to the heater, at the top and bottom.

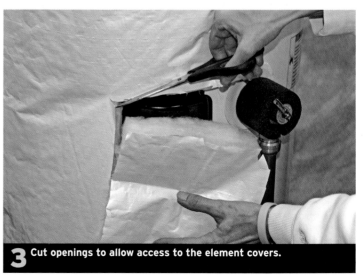

3 Cut openings to allow access to the element covers.

4 Wrap the blanket around the tank, keeping the top edge flush with the top of the heater.

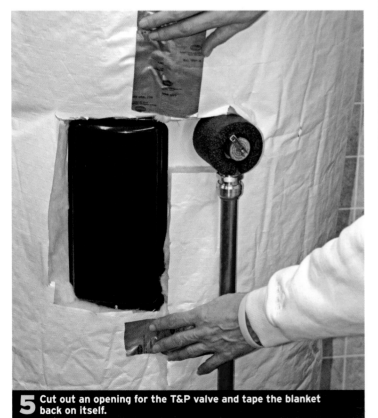

5 Cut out an opening for the T&P valve and tape the blanket back on itself.

6 Fold the rectangular blanket section in half and cut out holes for the water lines and the T&P valve, if it's mounted on top.

7 Tape down the corners and sides of the top section of the blanket.

WARNING

Do not cover the T&P valve or the element covers with an insulating jacket. Covering the valve may jam its lever, and you'll need access to the element cover for maintenance.

DRAINING AND FLUSHING A WATER HEATER

Whenever you need to change an element in an electric heater, leave the heater unused over a freezing winter, or when the heater needs cleaning, you'll need to drain it. You'll also need to drain it if it needs to be replaced. To drain a heater, you must open both its drain valve and the T&P valve (or a hot water faucet in the house) to break the vacuum the water will pull as it drains. Without breaking the vacuum, some water will always stay in the heater.

Before draining an electric water heater, turn off the electric power at a cutoff switch near the heater or at the circuit breaker or fuse panel. Throw the breaker (typically a 30-amp or 20-amp double pole) to the off position or remove both heater fuses ❶. For a gas heater, follow the manufacturer's shut-down procedures printed on the side of the tank. To keep water from coming back into the heater, turn the main supply valve off, too, normally at the valve on the cold side of the heater ❷.

Attach a garden hose to the drain valve on the bottom of the heater and pull the opposite end of the hose outside the house or push it into a floor drain ❸. The hose end must be level with or below the bottom of the heater to allow all the water to drain out.

Open the T&P valve by lifting the lever or turn on a hot water faucet anywhere in the house ❹. Then open the drain valve ❺. Let the heater drain completely, then close the drain valve and the T&P valve. When refilling, to make sure the air gets purged from the tank so it can fill completely, open a hot water faucet in the house (if you haven't already done so), and turn the faucet off when water begins to flow from it. Then turn off the water coming into the heater. Reapply the power to the heater.

To flush a water heater, first drain the heater. Leave the hose attached to the drain. Then turn the water heater valve back on in spurts (full on for a few seconds and then off) to dislodge as much debris as possible in the tank bottom. Do this at least a dozen times. Observe the drain hose end for muddy water and debris. Refill the heater when the hose runs clear.

TRADE SECRET

Hard water can build scale that reaches the bottom element in just a few months. If you don't drain the heater, the element will be covered by the scale and burn up.

1 Turn off the power to the heater by flipping the breaker to off or by removing the two fuses in the water heater electric circuit.

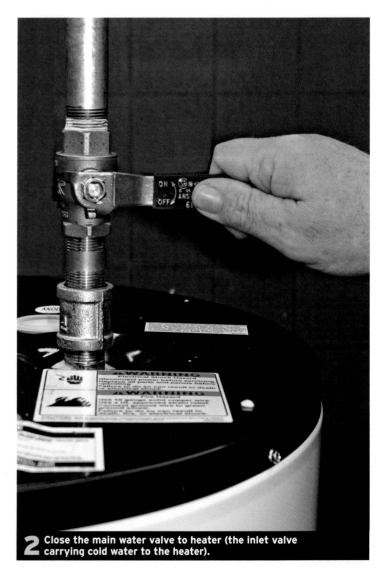

2 Close the main water valve to heater (the inlet valve carrying cold water to the heater).

LEAK CAUSES & CURES

Puddle of water around where pipe goes into heater. Could be caused by a leaking overhead pipe/fitting or connection itself to heater. Tighten connection if needed.

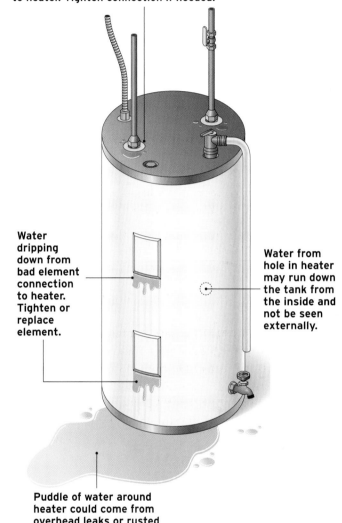

Water dripping down from bad element connection to heater. Tighten or replace element.

Water from hole in heater may run down the tank from the inside and not be seen externally.

Puddle of water around heater could come from overhead leaks or rusted hole in heater.

3 Attach a garden hose to the heater and pull the opposite end of the hose outdoors or insert it in a basement drain.

4 Break the vacuum by lifting the lever on the T&P valve.

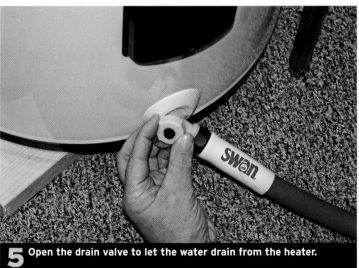

5 Open the drain valve to let the water drain from the heater.

⚠ WARNING

If there is a blockage (mud or iron deposits, for example), preventing the water from draining out the valve, you must unscrew the drain valve and stick in a screwdriver to punch a hole in the blockage. However, once done, the area around the heater will immediately flood with muddy water, so you'll need to prepare yourself for some cleanup.

REPLACING A TEMPERATURE AND PRESSURE VALVE

A temperature and pressure (T&P) relief valve prevents your water heater from becoming a bomb. If the inside temperature becomes too hot (because the thermostat malfunctions) or if the interior pressure becomes too great, the T&P valve provides the means for the water to escape. Typically, a pipe extends from the output of the valve straight down the side of the heater to about 6 in. off the floor. If the floor is finished, consider a drain pan and from there a drain pipe to a floor drain or to the outside. If the side discharge-pipe run is a straight line to the floor, you can make it up ahead of time and install it as a single assembly.

If you raise the valve lever and nothing happens (very hot water should come out), the valve is bad and needs replacing. Never test the valve without having a new one handy—once engaged, a bad valve may not shut off. If this happens, immediately turn off the cold-water valve and the power to the heater (if electric). Drain the water in the heater to a point below the T&P valve.

➡ See "Draining and Flushing a Water Heater," p. 198.

Remove the T&P valve pipe extension. If the valve is on top of the heater, the pipe will have to be cut (but note the direction it faces before you cut it) ❶. Using a large pipe wrench and a lot of muscle power, push down on the handle to turn the valve body counterclockwise. The larger the wrench, the easier it will be—and grabbing the end of the wrench will give you more leverage if you need it.

Turn the wrench until the valve comes loose enough for you to unscrew it by hand ❷. Clean the threads with a wire brush and rag and thread in the new valve by hand as far as possible ❸. Using the same pipe wrench, turn the valve until it is tight and pointing down. Be careful not to bend the female threads of the valve. For top-mounted valves, turn the valve to face the same direction as the old unit ❹. Screw the valve discharge pipe you removed into the threaded output of the new valve. Make the last turn with a wrench ❺. Placing a small piece of foam insulation over the valve top will insulate it and help keep the water warm ❻. Turn the water on and refill the tank, purging air from the system as you would after draining the tank for any purpose. Then turn the power on. If you decide to run the discharge pipe outside, be aware that the pipe cannot be run uphill—it must be pitched at least 1/8 in. per ft. downhill. Thus it cannot do loops and it cannot be decreased in size. Treat it as a common drain line to stay out of trouble. CPVC is allowed for this pipe.

1 Turn off the water and power to the heater, drain the water below the valve, and remove the discharge pipe.

4 Using the same pipe wrench, turn the valve clockwise until it's tight and the threaded output points down.

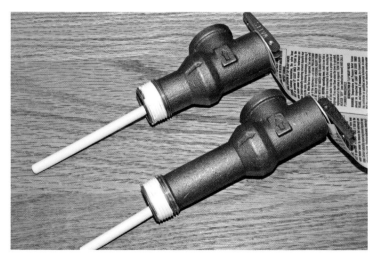

T&P valves come in short-shank models for heaters with standard insulation and long-shank models for heaters with extra-thick insulation.

2 Remove the valve by loosening it with a pipe wrench. Once it's loose, turn it by hand, and pull it away from the water heater.

3 Tape the threads of the new valve with Teflon tape, insert the valve into heater's threads, and tighten it hand tight.

5 Attach the valve discharge pipe to the threaded output of the valve.

6 Cut and insert a piece of foam insulation over the valve body.

Installing a CPVC discharge pipe

CPVC can be used for the T&P discharge pipe, and it's inexpensive. Install a brass male CPVC adapter in the valve. Glue a CPVC pipe into the male adapter. Cut the CPVC pipe to within 6 in. of the floor.

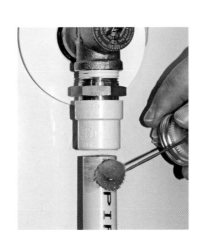

Reducing bushing   Threaded fittings on end of pipe

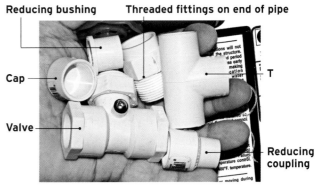

Cap

Valve

T

Reducing coupling

Codes won't allow these fittings in the discharge pipe because they reduce or interfere with pressure discharge.

REPLACING AN ELECTRIC WATER HEATER ELEMENT

Water heater elements heat the water in an electric water heater, and they can go bad with old age, exposure to hard or acid water, and being dry-fired–applying power to the heater with no water around the element. An upper element will be damaged if you don't purge the air out of the tank when refilling it because the air will prevent the water level from covering the element sufficiently. Older-style elements are bolted to the tank. New elements are screwed into threads in the tank body.

To remove an upper or lower element, shut off the power, turn off the water, and drain the heater.

➡ See "Draining and Flushing a Water Heater," p. 198.

Remove the element cover ❶. Do not lose the insulation (if any) that was under the cover plate. Next, you'll see a plastic snap-on cover that extends over the thermostat and element screws. Being careful not to break it, remove the plastic cover ❷. There will be two wires going to the element. Using a Phillips screwdriver, loosen each screw (do not remove them completely) and slip the wires out from under each screw ❸. Don't worry about which wire goes where. Then using a socket wrench and some muscle power, break the element free from the heater (it will be rusted solid), and unscrew it ❹. If you're replacing a bolted element, remove the bolts.

➡ See "Element-removal sockets," p. 203.

1 Unscrew the element cover screws and remove the cover.

If the element folds back on itself, it may not pull through the hole in the tank. Don't worry about damaging the element, it is bad anyway. Using some extra strength, pull hard and at angles to bend the element until it comes out ❺. Then screw or bolt the new element in place, reattach the wires, the plastic snap-on cover, and the cover plate. Turn on the water to check for leaks, and reapply the power.

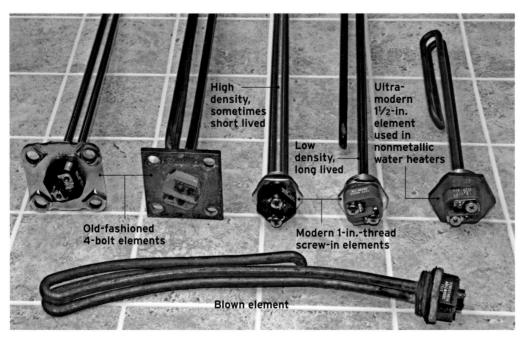

High density, sometimes short lived

Ultra-modern 1½-in. element used in nonmetallic water heaters

Low density, long lived

Old-fashioned 4-bolt elements

Modern 1-in.-thread screw-in elements

Blown element

Elements come in many different wattages, voltages, and sizes. Be sure to match your replacement to the old element. The bottom one was dry-fired and ruined.

⚠ WARNING

Elements vary in voltage and wattage. Match both on the new wattage on the new element to that of the old one–typically 240 volts at 4,500w for home heaters and 3,500w for some mobile home units. Look for the wattage and voltage stamped on the head of the element.

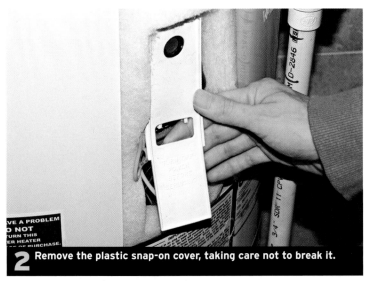

2 Remove the plastic snap-on cover, taking care not to break it.

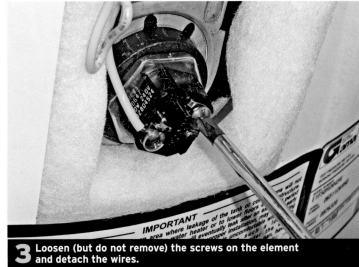

3 Loosen (but do not remove) the screws on the element and detach the wires.

4 Using a socket wrench, turn counterclockwise, and break the element free from the threads of the heater.

5 Once loose, slide the element out. Install a new element, wiring, snap-on cover, and metal cover plate. Turn the water and power on.

Element-removal sockets

To break rusted elements free from the tank, you can use a pressed metal socket (available at all hardware stores) but these tend to bend and slip off, and they won't give you leverage. You need all the leverage you can get, so use a 1½-in. socket with a ¾-in. breaker bar.

A low-cost socket works well for new elements but may not break an old element free of its threads.

Use a breaker bar and 1½-in. socket on rusted elements.

INSTALLING AN EXPANSION TANK

In the last several years, building codes have begun to require back-flow prevention valves in residential water systems. These valves prevent home water from flowing back into the municipal supply. But when the pressure of the municipal water is too high or when heating the water in your water heater overpressurizes the tank, the pressure can reach a dangerous level. An expansion tank installed in the cold water line to the water heater will counteract this problem. The tank has a flexible bladder that expands and contracts to keep pressure increases from damaging the glass liner of the water heater. Without this tank, increased pressure can trip the T&P valve or push out the bottom of a water heater like an egg.

Instructions that come with the expansion tank will require the tank to be installed on the cold side and pointing up or down. The best method is down and adjacent to the tank. When installing the galvanized fittings, be sure to wind the threads with Teflon tape.

Start by attaching a 3/4-in. galvanized elbow to the water heater. Then, to get the tank a few inches away from the heater, screw in a long metal nipple (around 12 in.) into the elbow ❶. Using a large pipe wrench, turn a 3/4-in. galvanized T-fitting on the long nipple and point it down ❷. Install an interface fitting for the type of pipe you're using—here, PEX pipe is shown with a universal fitting that will accept PEX, copper, and CPVC ❸.

> **See "Push-On Fittings," pp. 40–41.**

The expansion tank has 3/4-in. male threads. Turn the threads into the galvanized T-fitting ❹. Then determine whether or not you need to support the weight of the tank. If the outlet pipe from the tank turns up and is supported on the joists, you might not. But if the tank-outlet pipe continues horizontally, you'll need strapping fastened to the wall or down from the ceiling.

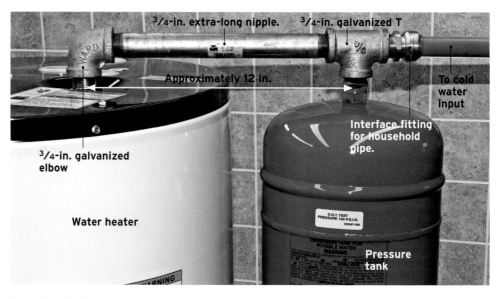

3/4-in. extra-long nipple.   3/4-in. galvanized T

Approximately 12 in.

To cold water input

Interface fitting for household pipe.

3/4-in. galvanized elbow

Water heater

D.O.T. TEST PRESSURE 120 P.S.I.G.

Pressure tank

Expansion tanks come in 2-gal. and 5-gal. sizes. The smaller size will serve the majority of residential water heaters.

1 Screw a 3/4-in. galvanized elbow on the nipple on the cold side. Add a 12-in. nipple.

2 Screw a galvanized T-fitting on the 12-in. nipple and point it down.

3 Attach an interface fitting that will accept the household pipe.

4 Attach the tank in the bottom opening of the galvanized T-fitting.

INSTALLING A FULL-FLOW WATER HEATER DRAIN

The drain of a common water heater is not of the highest design and quality. Its internal angles block sediment and make it impossible to get a wire through to break up blockages within the heater. As a remedy, you can remove the existing water heater drain and install a ball-valve with a hose-bibb connector. It's easier to replace the valve on a new heater before installation, but if you're replacing it on an existing heater, you'll have to drain the tank first.

➜ **See "Draining and Flushing a Water Heater," p. 198.**

Using an adjustable wrench, remove the existing drain ❶. Assemble a ³/₄-in. nipple, a ³/₄-in. ball valve, and a hose-bibb connector ❷. Be sure the handle of the ball valve extends away from the heater. Screw this assembly into the heater and tighten it, with the valve handle pointing up ❸. Turn the handle on and off to verify it will not hit the tank. Leave the handle off, 90 degrees to the pipe ❹.

INSTALLING AN EASY-DRAIN VALVE

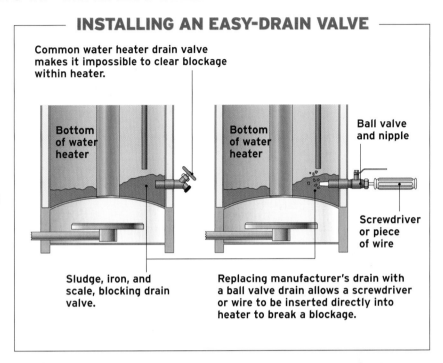

Common water heater drain valve makes it impossible to clear blockage within heater.

Bottom of water heater

Bottom of water heater

Ball valve and nipple

Screwdriver or piece of wire

Sludge, iron, and scale, blocking drain valve.

Replacing manufacturer's drain with a ball valve drain allows a screwdriver or wire to be inserted directly into heater to break a blockage.

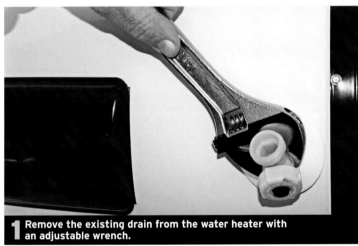

1 Remove the existing drain from the water heater with an adjustable wrench.

Male hose bibb by ³/₄-in. MIP

³/₄ in.-ball valve   ³/₄-in. nipple by 4 in.

2 Assemble the new drain assembly from a 4-in. nipple, a ball valve, and a hose-bibb connector. Tape all the male threads.

3 Install the new drain assembly, first hand tight, and then finish with a pipe wrench, leaving the handle on top of the valve.

4 Turn the handle to the off position and install the new tank or refill the existing tank.

REPLACING THE UPPER THERMOSTAT

The upper thermostat contains an overload switch (the red button) and an adjustable temperature control. At the factory, the thermostat is set at 120° F. But if the thermostat malfunctions and the water gets too hot, the sensor will remove power to the water heater. If you have no hot water, this red button is the first thing you test (by pushing it back in). If it continues to pop out, you need a new thermostat.

To replace the thermostat, first turn the power off. Then remove the top cover with a ¼-in. hex-head wrench or screwdriver. Set both the cover and screws aside ❶. Snap off the vertical plastic cover that isolates all the electrical screws ❷. Exposed, the upper section is the overload unit, the top four screws. Below that is the actual thermostat itself with a temperature adjustment. Notice the factory setting at 120°F ❸.

Make a sketch of the thermostat and log what color wires go where, then loosen the screws and pull the wires to the side with needle-nose pliers ❹,❺. The thermostat is held to the tank with two spring tension clips, on the bottom left and bottom right. Pull these out—first one side and then the other—with needle-nose pliers, while you slip the thermostat out from under them and away from the heater ❻,❼. The new thermostat should look exactly like the one you removed. Insert the new thermostat back under the clips and reattach the wires. Put the covers back on and reapply the power ❽.

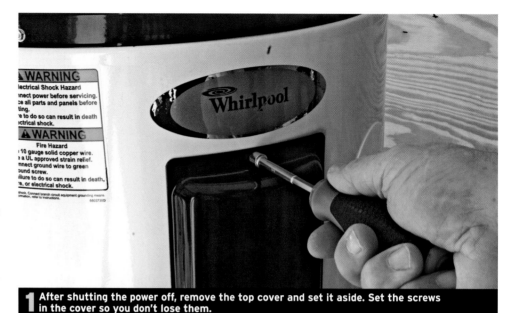

1 After shutting the power off, remove the top cover and set it aside. Set the screws in the cover so you don't lose them.

2 Removing the outside cover will expose a plastic cover. Snap it off its clips.

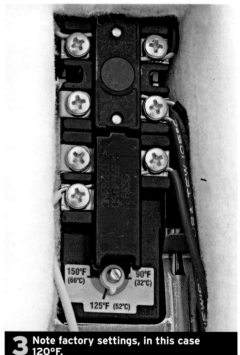

3 Note factory settings, in this case 120°F.

⚠ WARNING

Don't adjust the water heater thermostat above its factory setting of 120°F. Water at temperatures above 120°F can scald you, and it's easy to break a bone falling in the tub as you try to avoid the hot water if you accidentally turn the wrong knob.

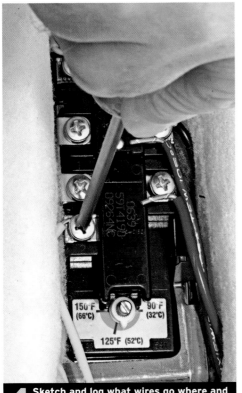

4 Sketch and log what wires go where and loosen the wiring screws.

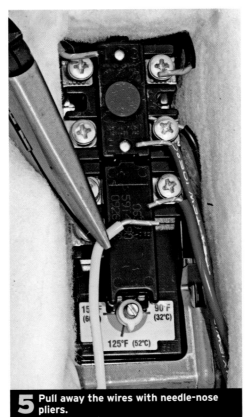

5 Pull away the wires with needle-nose pliers.

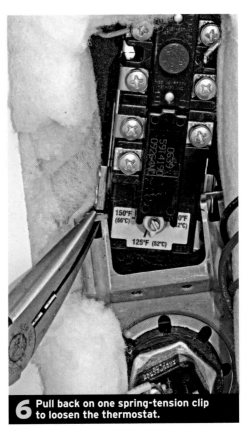

6 Pull back on one spring-tension clip to loosen the thermostat.

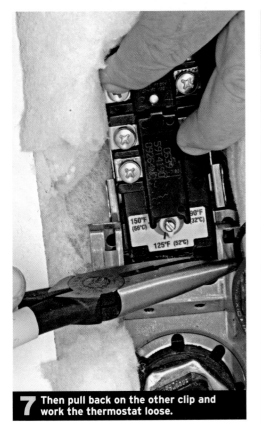

7 Then pull back on the other clip and work the thermostat loose.

8 Remove the thermostat from the heater and replace it with an identical unit.

Replacing the lower thermostat

Use the instructions for replacing the upper thermostat. The bottom thermostat will have fewer wires and will be faster and easier to replace.

The bottom thermostat only has two wires, making replacement easy.

REPLACING AN ANODE ROD

The anode rod in an electric water heater attracts dissolved contaminants that would otherwise rust out the interior of the tank. In the process, the anode rod will corrode, but as soon as the rod is eaten away, the water will start to corrode the tank. It's important, therefore, to change the rod once it's eaten up by the water. Unfortunately, there is no way to know the condition of the rod unless you pull it. A good rule of thumb is to check it approximately every 5 years.

Before you pull the rod, turn the power off, and turn the main water valve off, as well. Let the water cool for a couple hours, then run some cold water through the tank so you don't get burned.

Sometimes the anode rod's socket head can be seen at the top of the heater and other times it may be hidden under insulation. In the latter case, you will have to cut the insulation away ❶. Using a 1¹/₁₆-in. socket on a ³/₄-in. breaker bar, turn the bar counterclockwise. The rod will be very hard to turn. It may take two people pulling on the breaker bar while someone may have to hold the heater to keep it from turning and breaking the supply-line connections ❷. Once loose, pull the rod straight out. The rod will be around 3 ft. to 4 ft. long. If it's eaten down to its center wire, replace it ❸. Slide the new rod into the opening, tighten it, and turn the water and power back on. Segmented rods are available online for rooms with limited overhead clearance (the rods bend at the segments, allowing you to clear the over-head framing as you pull it out). For water with a high sulfur content, you can also order special anode rods that will remove the rotten-egg smell from the water.

1 Find the anode rod socket head at the tank top, scraping away any insulation, if necessary.

2 Using a 1¹/₁₆-in. socket on a ³/₄-in. breaker bar, unscrew the anode rod. You may have to hold the heater.

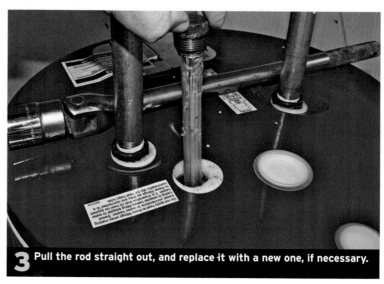

3 Pull the rod straight out, and replace it with a new one, if necessary.

⚠ WHAT CAN GO WRONG

Anode rods are not stocked by suppliers like they used to be. Thus locate a source and get the rod ahead of time. You may have to get it off the internet or have it special ordered by a supplier.

INSTALLING A RECIRCULATOR

In a typical water system, there's at least one faucet whose distance from the water heater requires running a significant amount of cold water down the drain until the hot water reaches the fixture. This practice is extremely wasteful, but with a recirculator, it becomes unnecessary. Many current recirculators are elaborate systems for basement installation. Some don't work well, and some need professional installation. However, at least one unit—the RedyTemp® recirculator—can be installed by the homeowner at a bath or kitchen sink, and it takes only a few minutes. This system employs a thermostat to sense cooled water in the hot line and automatically sends it back into the cold line. Not only do you have hot water ready at the faucet, but you don't waste (and pay for) discarded cold water. The RedyTemp can also be used with a timer. Certain models add a variety of other optional sensors.

Before you start, purchase two supply tubes with 1/2-in. female threads on both ends (if they're not already included with the unit). Slide the recirculator under the sink, typically the sink most distant from the water heater. Turn off both stop valves and loosen and remove the supply tubes from the faucet ❶. Connect these ends of the supply tubes to the two male connections at the rear of the circulator—hot on the left, cold on the right ❷. Then connect one end of your purchased supply tubes to the two front nipples on the recirculator ❸. Connect the other end of the supply tubes to the faucet, with the left-front tube going to the hot water faucet and the right-front tube to the cold water faucet. Be sure all the connections are tight at both ends. Then turn the water back on at the stop valves. Drill a 1 1/8-in. hole in the side of the sink cabinet that's closest to a GFCI outlet. Plug the unit into the outlet, using timers as specified by the manufacturer. If drilling a hole through the cabinet is not desirable, then install a GFCI outlet under the sink ❹. If you already have two outlets under the sink, plug the unit into the dishwasher outlet, not the disposal outlet.

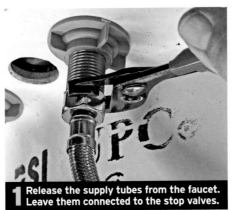

1 Release the supply tubes from the faucet. Leave them connected to the stop valves.

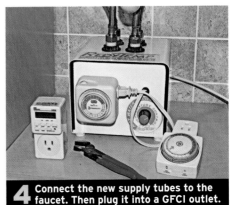

2 Screw the tubes from the stop valves to the rear nipples on the recirculator.

3 Connect one end of the two new supply tubes to the front two nipples.

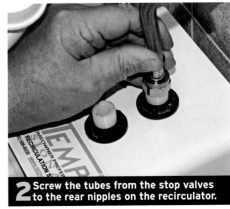

4 Connect the new supply tubes to the faucet. Then plug it into a GFCI outlet.

RUNNING WATER LINES TO A RECIRCULATOR

The circulator (RedyTemp type) should be installed at the end of the plumbing run. Typically this is the bathroom sink or the kitchen sink.

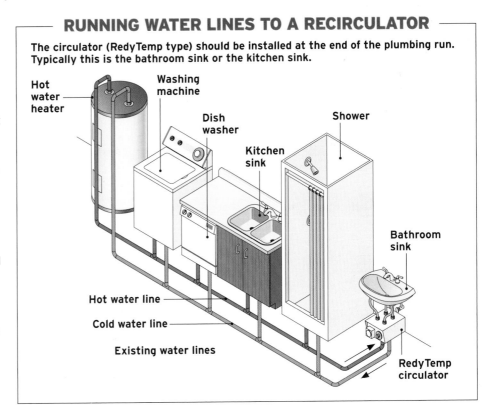

Hot water heater

Washing machine

Dish washer

Kitchen sink

Shower

Bathroom sink

Hot water line

Cold water line

Existing water lines

RedyTemp circulator

INSTALLING A WATER SOFTENER

If you suspect you have hard water or iron in your system, you can have your water tested. Do-it-yourself kits are available, but the results are often difficult to interpret. You can also have your water tested by a retailer, which is usually a free service. However, you normally know if you have problems without any test. A white coating on a shower wall and on faucets (from hard water) indicates the presence of mineral residue in the water, and you need a softener to keep this off your body, out of your hair, and out of your clothes. Iron deposits also leave telltale signs. You will have iron-stained clothes, tubs, and showers, and your water will taste like iron. Tap off the untreated water side for outside spigots.

Sizing a water softener

Testing by a water softener retailer will give you a readout of the grains of hardness in your water, and that result will automatically size the softener. Typically, for a family of four, if your water tests at less than 10 grains of hardness, you can get by with a 20,000-grain unit. Above 10 grains but below 20, use a 30,000-grain unit. Above 20 but below 30, a 50,000-grain unit. As an added bonus, water softeners will remove clear-water iron. Indeed, water softeners are sometimes installed just to remove the iron, if the amount is small.

Installing the softener

Unpack the softener close to its final location, and read all the directions. You'll have to shut off the water to the house and cut into the existing water line close to where it enters the house.

➡ See "Repairing and Tapping Water Lines," p. 44.

Though the installation procedures for a softener may vary by manufacturer, the general procedures are the same for all of them.

WARNING

If you cut a metal water line to install a softener, and that softener has a plastic head, you will need an electrical bypass (a wire) from the metal pipe to the softener body to maintain continuity.

If your glass shower walls **look like this, you need a softener.**

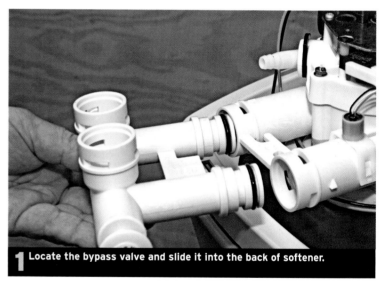
1 Locate the bypass valve and slide it into the back of softener.

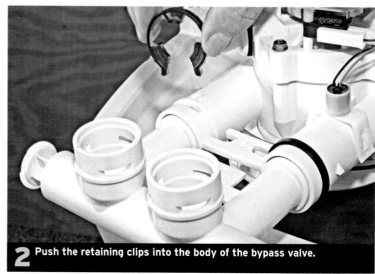
2 Push the retaining clips into the body of the bypass valve.

First, locate the bypass valve unit and slide it into the back of the softener. Do not damage the O-rings **❶**. To keep the water pressure from blowing the valve back out, snap in the two retaining rings **❷**. Push the drain hose all the way on the barbed outlet on the head of the softener. If the hose is stiff, warm it for a couple of minutes with a hair dryer. A wire spring clip is usually provided to snap around the hose to keep it tight on the outlet. These are very hard to compress and keep straight on the pipe, so installers commonly use a small hose clamp instead **❸**. Install the overflow tube on the overflow elbow, which is located on the side of the unit **❹**. This is the outlet that lets an overfilled tank release excess salt water, and ultimately you'll have to connect it to a hose that runs to the same drain as the high-pressure drain that comes off the softener head.

Insert the two interface connections into the softener head, and lock them in place with the manufacturer's clamps **❺**. Next install fittings on the softener head appropriate to the kind of supply pipe you're using. The exact fitting you use will depend on the type of pipe you have in your house. The fitting will need to have 1-in. female threads on one end (assuming the softener has 1-in. threads, as most do) and a pipe interface thread or adapter on the other.

➡ **See "Water-Line Interfaces," p. 212.**

Set the softener exactly where you want it to be—preferably against a wall where the main water line is located, and if you are crimping PEX fittings, make sure there is room to crimp.

>> >> >>

Water softener checklist
Before you go out and buy that softener, make sure you have:
- **A 120v outlet for electrical power.**
- **An inside house drain if not drained outside.**
- **Room—a softener takes a large footprint, at least 3 ft. left to right.**
- **Heat—you cannot locate a softener in an area that will get below freezing.**
- **A location close to where the water enters the house.**

WHAT CAN GO WRONG
If the house main valve is a gate valve (round handle), there is a possibility that the water will not fully turn off—a dribble will come through. It is not fixable. You will either have to live with it or change the valve out to a ball valve (lever handle).

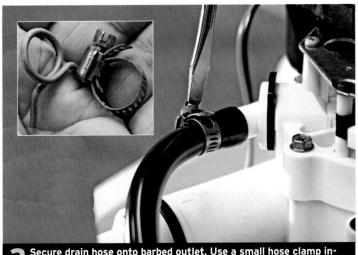

3 Secure drain hose onto barbed outlet. Use a small hose clamp instead of the spring clamp supplied by the manufacturer (see inset).

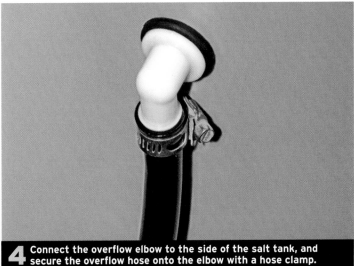

4 Connect the overflow elbow to the side of the salt tank, and secure the overflow hose onto the elbow with a hose clamp.

5 Insert the pipe clip-on fittings, and secure them with the snap rings supplied by the manufacturer.

INSTALLING A WATER SOFTENER (CONTINUED)

Always place the unit before you pour in the salt because the salt will make the unit a lot heavier. Then install the water lines to the interfaces. Pour in several bags of the salt recommended by the manufacturer, along with a couple gallons of water. For water with heavy iron, consider using a special salt developed for heavy-iron water. Plug the softener into an AC outlet, and program its controls according to the manufacturer's instructions.

WATER LINE INTERFACES

Install the interface fittings onto the clip-on fittings. For PEX, use a 1-in. female thread union by a $^3/_4$-in. crimp interface fitting. For CPVC, use a 1-in., CPVC female thread fitting (with gasket) with a $^3/_4$-in. glue joint and a 1-in. by $^3/_4$-in. reducing bushing. For a universal connection that fits copper, PEX, and CPVC, use a 1-in. by $^3/_4$-in. galvanized reducing coupling along with a $^3/_4$-in. male thread Sharkbite (push-on) adapter. For copper, use a 1-in. female thread by $^3/_4$-in. sweat connection for about 1 ft. of copper pipe. Above that, use a push-on coupling to connect to the existing copper pipe.

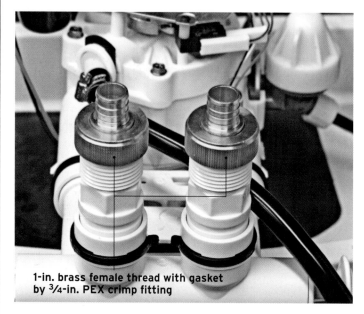

1-in. brass female thread with gasket by $^3/_4$-in. PEX crimp fitting

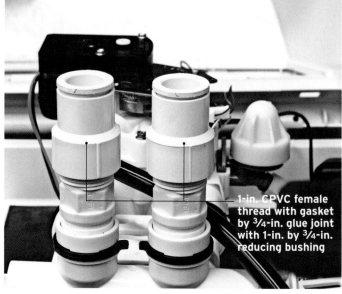

1-in. CPVC female thread with gasket by $^3/_4$-in. glue joint with 1-in. by $^3/_4$-in. reducing bushing

$^3/_4$-in. brass universal push-on male adapter

1-in. by $^3/_4$-in. galvanized reducing coupling

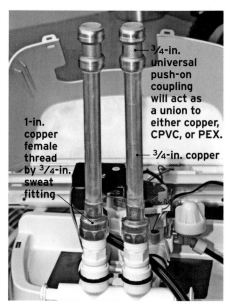

$^3/_4$-in. universal push-on coupling will act as a union to either copper, CPVC, or PEX.

1-in. copper female thread by $^3/_4$-in. sweat fitting

$^3/_4$-in. copper

INSTALLING A BASEMENT REVERSE-OSMOSIS FILTER

1 Mount a piece of plywood to the wall and drive in two screws spaced to hold the module.

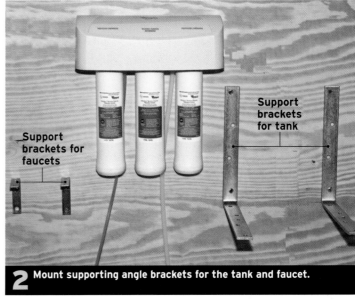

2 Mount supporting angle brackets for the tank and faucet.

Support brackets for faucets

Support brackets for tank

3 Drill a 1¼-in. hole in a 2x4 block and cut a key for the pipes. Mount the faucet to the block and screw it to the brackets.

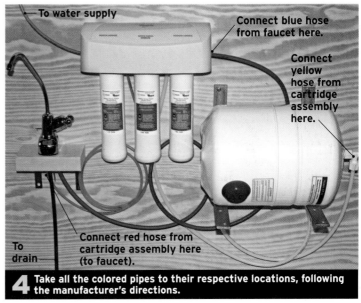

To water supply

Connect blue hose from faucet here.

Connect yellow hose from cartridge assembly here.

To drain

Connect red hose from cartridge assembly here (to faucet).

4 Take all the colored pipes to their respective locations, following the manufacturer's directions.

Installing a reverse osmosis system in the basement instead of under the kitchen sink frees up storage space under the sink and lets you use existing basement drains instead of cutting into the kitchen drain.

> **See "Installing a Reverse-Osmosis Filter," pp. 173-175.**

You can mount the system directly to a block wall or to a plywood backing. You'll need a pistol-grip hammer drill, a concrete bit, and concrete screws to install the system on a masonry wall. A plywood backing, however, reduces the number of holes you have to drill in concrete or block, making the job easier.

No matter what method you use, first mount the filter unit on screws spaced as specified by the manufacturer **1**. Mount angle brackets—two large ones for the tank and two small brackets for the faucet support. Bend the tank brackets to form a slight V and set the tank in this "cradle," secured with large plastic ties **2**. Mount the faucet to a 2x wooden block after cutting a 1¼-in. hole and a "key" in the back for the pipes. Attach the faucet block to the small angle brackets **3**. Run the colored lines to their respective locations, following the manufacturer's instructions **4**.

INSTALLING A CARTRIDGE FILTER

You need a cartridge filter if there is debris or iron in your water. A cartridge filter will not replace a full-sized water conditioner, but if the iron level is not excessive, and you don't mind changing the filter often, a cartridge filter is a low-cost alternative.

The three rules for installing a cartridge filter are secure, secure, secure. The mounting has to be strong enough to resist the pressure of the large wrench you'll use when you remove the housing to change the filter. The best and least costly mounting is a combination of angle brackets and U-bolts.

First, fasten the brackets to the wall close to the service-pipe entry and about 8 in. apart. For concrete or block walls, use masonry screws ❶.

➜ **See "Drilling Masonry Walls," p. 216.**

Then, using one of the predrilled holes in the brackets, mark and drill a second hole in each bracket at the width of the U-bolt. Be sure the holes are far enough away from the wall to allow room for the filter and the filter wrench ❷. Attach a 3/4-in. nipple to each side of the filter. Use nipples long enough to extend beyond the brackets, generally 6-in. nipples will do. Galvanized nipples can be painted to match the wall, but you can go fancy with brass ❸. Set the filter body and nipples on the brackets, slide the U-bolts over the nipples into the holes, and tighten the U-bolt nuts ❹.

Attach female fittings to the nipples, using fittings that will interface with the house pipe. The easiest method is to use a universal brass Sharkbite fitting that will interface with copper, PEX, or CPVC ❺. To be pipe specific, you will have to sweat if you want copper, crimp if you want PEX, and glue if you want CPVC.

➜ **See "Filter Interfaces," p. 215.**

➜ **See "Push-On Fittings," pp. 40-41.**

To change the filter, turn the bypass valve off, unscrew the housing, replace the filter and housing, and turn the valve back on.

1 Secure the angle brackets to the wall.

2 Drill hole(s) in the angle brackets spaced to fit a U-bolt.

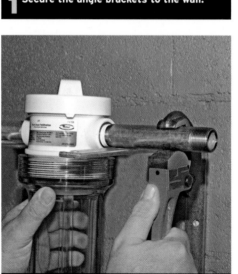

3 Attach galvanized nipples to both sides of the filter.

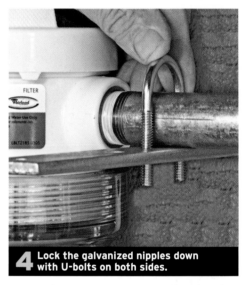

4 Lock the galvanized nipples down with U-bolts on both sides.

5 Install the appropriate fitting to interface with the house pipe. A female universal fitting, such as this Sharkbite push-on fitting, will allow easy interface with any pipe.

Picking a cartridge filter

Cartridge filters come in different sizes but none come without their pros and cons. Large-diameter filters don't need changing as often, but replacement filters are often hard to find, and you have to add a ball valve on each side because they don't come with bypass valves. When possible, get a filter that has an off, on, and bypass valve on its head. This valve will prevent water from coming through the filter outlet when you change it. Be sure to buy a filter with metal threads for the nipples.

The most common filter comes in a diameter of about 4 in. Always choose a clear housing so you can see how dirty the filter is.

For excessively dirty water or to allow a longer period of time between changes, use a large filter.

FILTER INTERFACES

Once you've mounted the cartridge filter and installed the nipples, you'll have to provide an interface from the nipples to the water pipe used in your home. The easiest method is to use a universal push-on fitting that will accept copper, PEX, or CPVC. You can, however, install transitions that are pipe specific, using one of the methods shown here.

For copper pipe, use a copper female fitting and sweat the copper pipe on it.

For PEX, use a female PEX crimp adapter fitting.

For CPVC, use a brass female glue-on fitting.

PLUMBING A WASHER

The simplest and least expensive method to get water to a clothes washer is with hose bibbs screwed into drop-ear elbows. Drop-ear elbows are made for all pipes. Pick the elbow for the type of pipe you are using—PEX, CPVC, or copper—and secure it to the wall ❶. Screw the hose bibb in the drop-ear elbow, first hand tight, then wrench tight, leaving the hose threads pointing straight down ❷. You can also use a boiler drain or a ball-valve boiler drain. Basically any valve with ½-in. male threads and a hose bibb on the opposite end will work.

1 Install a drop-ear elbow that will accept the water-line you are using.

2 Screw the hose bibb into the drop-ear elbow, with the hose threads down.

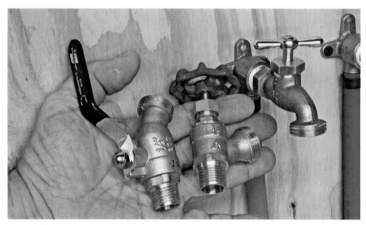
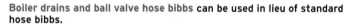

Boiler drains and ball valve hose bibbs can be used in lieu of standard hose bibbs.

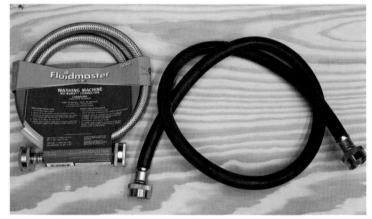

Common black washer hose has a habit of blistering and leaking. Use heavy-duty stainless-steel braided hose instead.

DRILLING MASONRY WALLS

Drilling a masonry wall requires a few special tools. A hammer drill does the trick, because it hammers as it turns, making the work go quickly. Buy masonry screws (with ¼-in. or ⁵⁄₁₆-in. hex heads) by the box, and the correct drill size will come with it.

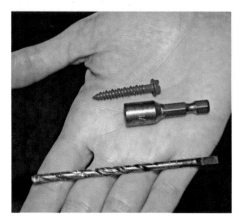

To drill concrete, you need a hammer drill, a carbide bit, and concrete screws.

Drill holes at the appropriate locations with a carbide bit.

Anchor the fixture with concrete screws, using a hex drive on your cordless drill.

A Good. This box has easy-turn valves.

B Better. This box has a single handle for both hot and cold.

C Best. This box has a single easy-turn handle and anti-shock valves.

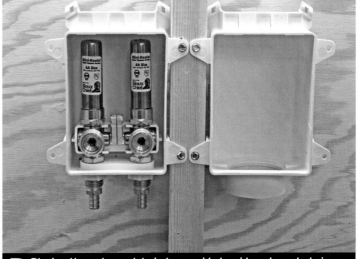

D Placing the water and drain boxes side by side onto a stud gives you the option of putting the drainpipe in a different stud cavity.

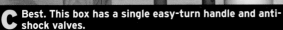

Picking a washer box

Washer boxes are typically designed to mount between studs (or on a stud) using supplied brackets. Don't buy a washer-box-only kit. It's cheap and not a true kit. This kind of box requires long-shanked valves (the ½-in. threaded ends must be long enough to fit through the box and still have enough length for the water line), and such valves generally have to be special ordered.

The valves within the boxes define the quality of the unit. A good box will have valves with lever handles **A**. A better box will feature a single easy-turn lever that closes and opens both valves at the same time **B**. The best box will use a single easy-turn lever but will also incorporate antishock valves to cushion the water hammer vibration that occurs when the washer solenoid turns on and off **C**. Installing the water lines and drains on either side of a stud offers you optional drain installations and a neater-looking setup **D**.

WARNING
Avoid washer boxes with round-handled valves. In time, these valves corrode in place and become impossible to turn.

INSTALLING A WASHER DRAIN

A washer drain consists of 2-in. drainpipe (typically schedule 40 PVC) with a stand pipe of around 36 in. that feeds into a 2-in. horizontal P-trap **Ⓐ**,**Ⓑ**,**Ⓒ**.

WHAT CAN GO WRONG

Codes require a horizontal run after the trap of at least twice the pipe diameter, or 4 in. for a 2-in. drain. This horizontal run keeps the P-trap from functioning as an unintentional and illegal S-trap. The horizontal distance reference is from the trap weir (the water in the trap) to the bend of the elbow or T-fitting, not the length of horizontal pipe between the two.

WASHER DRAIN SPECIFICATIONS

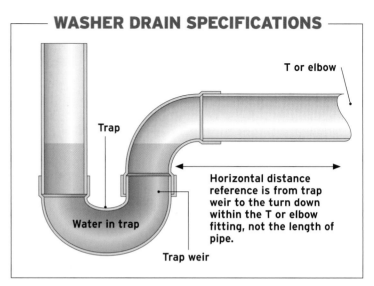

T or elbow

Trap

Water in trap

Horizontal distance reference is from trap weir to the turn down within the T or elbow fitting, not the length of pipe.

Trap weir

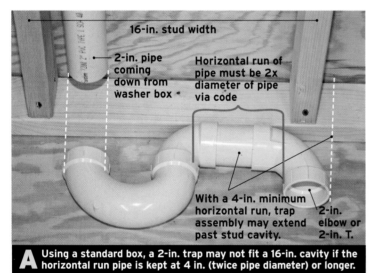

16-in. stud width

2-in. pipe coming down from washer box

Horizontal run of pipe must be 2x diameter of pipe via code

With a 4-in. minimum horizontal run, trap assembly may extend past stud cavity.

2-in. elbow or 2-in. T.

A Using a standard box, a 2-in. trap may not fit a 16-in. cavity if the horizontal run pipe is kept at 4 in. (twice pipe diameter) or longer.

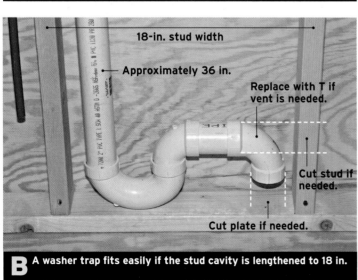

18-in. stud width

Approximately 36 in.

Replace with T if vent is needed.

Cut stud if needed.

Cut plate if needed.

B A washer trap fits easily if the stud cavity is lengthened to 18 in.

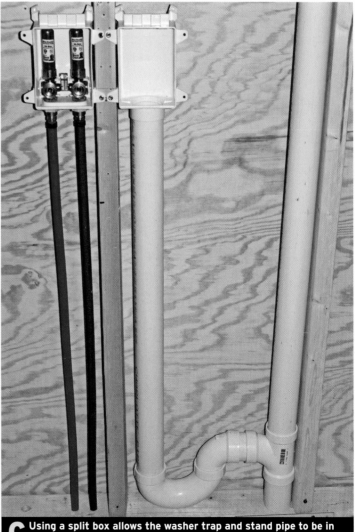

C Using a split box allows the washer trap and stand pipe to be in the same wall cavity without interference from the water lines.

HOOKING UP A WASHER

Once the valves and drain are in place, it's time to hook up the washer. This means connecting the washer hoses to the washer and the valves in the washer box as well as connecting the washer drain.

Just before you push the washer in place, tighten the washer hoses on the washer solenoid ❶. Attach the opposite end of the washer hoses to the valves in the washer box ❷. Be sure to keep hot to hot and cold to cold when you attach the hose ends to the valves in the box. Then insert the end of the washer drain hose into the drainpipe of the washer box ❸. Turn the water on, and check for leaks. To verify hot is hot and cold is cold, turn the washer to cold wash and check that cold water is going into the tub.

WHAT CAN GO WRONG

In new construction, most codes require washer outlets to be roughed in, even if the owner doesn't want them.

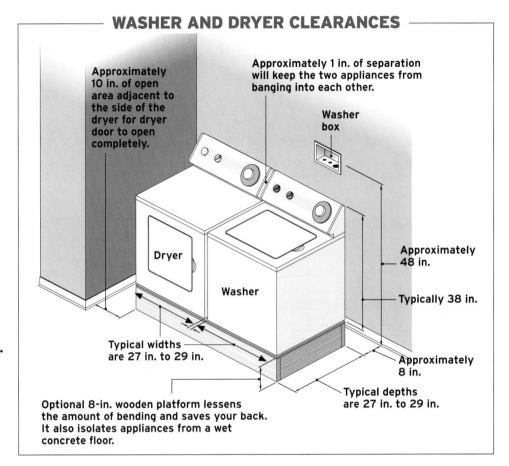

WASHER AND DRYER CLEARANCES

Approximately 10 in. of open area adjacent to the side of the dryer for dryer door to open completely.

Approximately 1 in. of separation will keep the two appliances from banging into each other.

Washer box

Dryer

Washer

Approximately 48 in.

Typically 38 in.

Approximately 8 in.

Typical widths are 27 in. to 29 in.

Typical depths are 27 in. to 29 in.

Optional 8-in. wooden platform lessens the amount of bending and saves your back. It also isolates appliances from a wet concrete floor.

1 Attach the hoses to the washer outlet on the solenoid.

2 Attach the other end of the washer hoses to the valves in the washer box.

3 Insert the washer drain into the drain pipe.

INSTALLING A UTILITY SINK

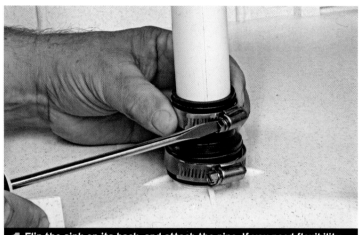

1 Flip the sink on its back, and attach the pipe. If you need flexibility, install a flexible trap adapter, then a short thin-wall pipe.

2 Insert the legs in the slots in the sink base, pushing them by hand in as far as they will go.

Back in the days of our fathers and grandfathers, utility sinks were cast from concrete, both single and double bowls. Needless to say they were very heavy. Today's sinks are made from plastic, and one person can install them without any problem. Their one drawback is that their light weight makes them unstable. That's why it's smart to fasten the back vertical lip of the sink to the wall behind it.

Any bath faucet with a swinging spout and 1/2-in. shafts spaced on 4-in. centers will fit a utility sink. However, no model except a utility faucet will have hose threads on the end spout, and the ability to attach a hose will increase the usefulness of your sink. Hose-bibbs with aerators are available.

Start by flipping the sink upside down on the floor. You can attach a short pipe with a gasket on the drain threads, but using a flexible trap adapter with a straight pipe will allow you to "cheat" a little if everything doesn't fit perfectly **1**. Slip the four legs into their slots in the sink corners **2**, then set them the final 1/2 in. with a rubber mallet. If

they still have a tendency to fall out, predrill a hole and drive in a screw to lock the legs **3**.

Typically, you'll want to center the faucet on the sink. Look for pressed circles which designate the normal location of the faucet holes. Measure the center-to-center shaft distance, and mark your hole locations **4**.

Using a hole saw slightly larger than the faucet shanks, drill out the two shank holes **5**. If the faucet body has no gasket underneath, caulk around the holes, and around the edge of the faucet **6**. Set the faucet in the holes and, working from underneath, lock the faucet down with the two nuts supplied **7**. Attach flexible supply tubes to the faucet's shanks, and connect the opposite ends to the stop valves under the sink **8**. Screw the lip of the sink to the wall through predrilled holes **9**. Connect the tub drain to a P-trap and the P-trap to the house drain. If you are using a dual-tub sink, connect the two drains with the supplied drainpipes **10**. Then add the tail piece and the P-trap.

Utility sinks come in **double- and single-bowl models. They're lightweight and easy to install.**

If you don't need a flexible trap adaptor, **install a straight pipe with a gasket.**

Gasket

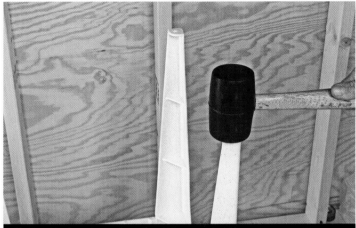

3 Set the legs in their slots with a couple of taps from a rubber mallet. Fasten them with predrilled screws for greater stability.

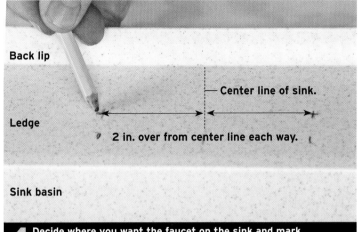

Back lip

Ledge

Sink basin

Center line of sink.

2 in. over from center line each way.

4 Decide where you want the faucet on the sink and mark the drain hole centers with a soft pencil.

5 Drill out the holes for the faucet shanks using a hole saw slightly larger than the shank diameter.

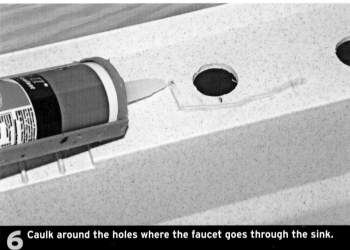

6 Caulk around the holes where the faucet goes through the sink.

7 Place the faucet in the sink and secure it from underneath by tightening the faucet mounting nuts.

8 Attach supply tubes to the faucet shanks and to the stop valves at the other end.

INSTALLING A UTILITY SINK (CONTINUED)

9 Secure the tub to the wall with predrilled screws through the back lip.

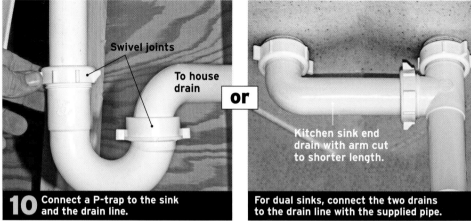

Swivel joints

To house drain

or

Kitchen sink end drain with arm cut to shorter length.

10 Connect a P-trap to the sink and the drain line.

For dual sinks, connect the two drains to the drain line with the supplied pipe.

SUMP PUMPS

When choosing a sump pump, you have two choices: pedestal or submersible. Always opt for the latter; they are smaller, easier to install, and require less maintenance. If you are worried about power outages, then you need a pump with a battery backup. This unit runs off computerized controls from a car battery. When replacing a sump pump, simply swap one pump out for another—submersible for submersible or pedestal for pedestal, or whatever is the easiest.

Your biggest decision when installing a sump pump will be where to dig the hole in the basement floor. To punch a hole in the concrete, you'll need a heavy-duty rotary hammer, which you can get from a tool renting company. Make sure you put the hole at the lowest place in the flooding area or exactly where the water comes in.

Choosing a sump pump
The gallons per hour (gph) rate will determine the size of pump you need, along with how high it has to be pumped. Typically a $1/3$-HP pump or a $1/2$-HP pump is used. You don't need more horsepower unless you are pumping higher than the basement ceiling;

then opt for a $3/4$- or 1-HP model. The gph rate is determined by how many gallons you estimate are flooding into your basement.

Installing a sump pump
Screw a male adapter with Teflon-taped threads into the sump pump's outlet, typically $1\frac{1}{2}$ in. PVC, but 2 in. for large pumps and $1\frac{1}{4}$ in. for small models. Glue a 2-ft. piece of pipe into the male adapter. Following the manufacturer's directions, attach the check valve, (glue-on or slip-on) onto the top of the pipe and make sure the check valve is pointing the right direction. This is one assembly, and it will be placed into the sump tank as a unit. Place the pump in the hole in the floor, and run a pipe (from the check valve) vertically and horizontally as needed to your proposed exit point on an exterior wall. Drill the exit hole in the rim joist at the top of the wall or cut a hole in a concrete wall with the rotary hammer.

A submersible sump pump can be totally submerged in the water. (Photo courtesy of Ridge Tool Company)

A pedestal pump is the old standard—pump on the bottom, motor on top. (Photo courtesy of Ridge Tool Company)

TRADE SECRET

Sump pumps need a check valve to keep the water from flowing backward once the pump turns off and a 1/4-in. weep hole in the pipe below the check valve to keep the system from pulling a vapor lock.

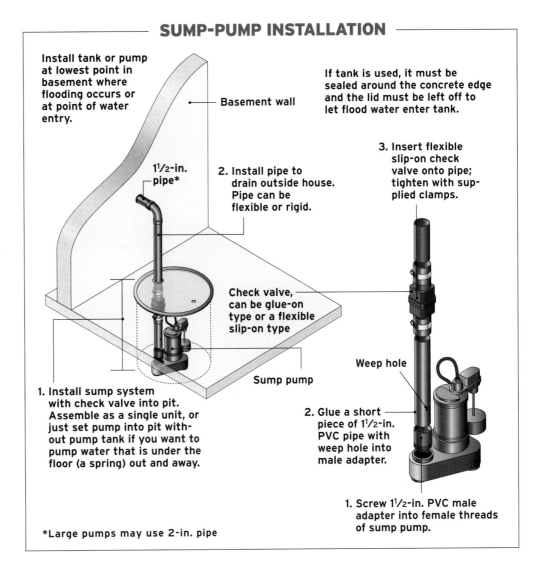

SUMP-PUMP INSTALLATION

Install tank or pump at lowest point in basement where flooding occurs or at point of water entry.

Basement wall

If tank is used, it must be sealed around the concrete edge and the lid must be left off to let flood water enter tank.

1½-in. pipe*

2. Install pipe to drain outside house. Pipe can be flexible or rigid.

3. Insert flexible slip-on check valve onto pipe; tighten with supplied clamps.

Check valve, can be glue-on type or a flexible slip-on type

Sump pump

Weep hole

1. Install sump system with check valve into pit. Assemble as a single unit, or just set pump into pit without pump tank if you want to pump water that is under the floor (a spring) out and away.

2. Glue a short piece of 1½-in. PVC pipe with weep hole into male adapter.

1. Screw 1½-in. PVC male adapter into female threads of sump pump.

*Large pumps may use 2-in. pipe

The battery backup DC pump system can work even when there is a power outage. (Photo courtesy Basement Watchdog Systems)

INSTALLING AN UP-FLUSH SYSTEM

Installing a toilet or full bath in your basement will add convenience to your home and can increase its selling price. A basement toilet or bath installation, however, will prove difficult if you have to cut the concrete floor to accommodate the drain line.

As an easier alternative, you can install an "up-flush" system. This system consists of a tank that sits on the floor (no digging required) with a sewage pump inside the back. The pump pushes waste water up a drainpipe via a fitting cut into the main drain overhead. The unit will accommodate a toilet and a tub/shower and sink. Its only disadvantage is that it puts a toilet about 5½ in. higher and a tub or shower 6 in. to 7 in. higher than if they were installed on the existing floor. But you can get around this problem by building a false floor to the same level.

You'll have to cut the 2-in. pressurized discharge pipe from the sewage pump into the drain lines overhead, but with a flexible fitting, that takes only a few minutes. Another flexible fitting will transition the 3-in. vent pipe from the tank into the house vent system or outside through the rim joist or the top of a masonry wall. You cannot reduce the vent size.

You'll find the easiest installation is with a Zoeller® Qwik-Jon® kit (available at most large plumbing supply firms), which has all the parts in one package. Allow all day for installation, and make sure no one uses the drain/sewage system when you have lines cut. Read and understand all directions before you start.

When you get the kit home, verify all parts are in the kit, and get familiar with them. You will note the tank is flat in front with a raised compartment (the "tower") in the back. The flat section supports the toilet, and the tower encloses the sewage pump.

Now comes one large decision—you'll have to decide whether to install the tower in the room or behind a wall ❶.

>> See "Choosing the Tower Location," on the facing page.

To add drains for a bath/shower or sink, drill into the side of the tank only where indicated by manufacturer. Insert the supplied grommet for a 2-in. pipe in the hole ❷,❸.

A shower or tub will need to be on a raised platform slightly higher than the drilled hole so the water can flow downhill to the tank.

>> >> >>

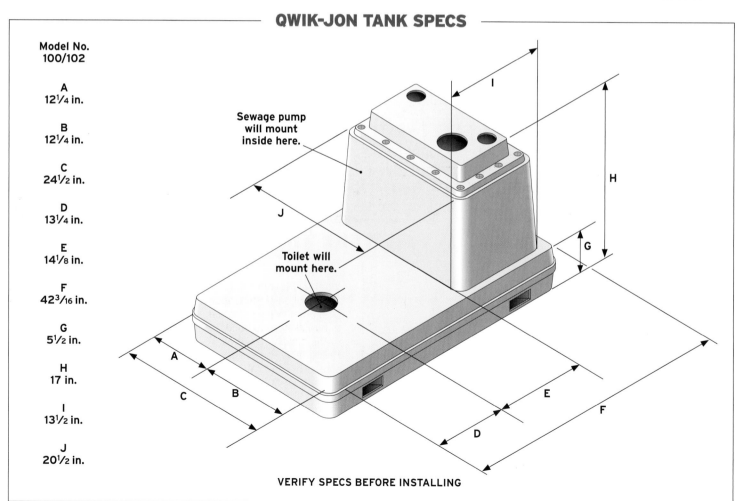

QWIK-JON TANK SPECS

Model No. 100/102

A 12¼ in.

B 12¼ in.

C 24½ in.

D 13¼ in.

E 14⅛ in.

F 42³⁄₁₆ in.

G 5½ in.

H 17 in.

I 13½ in.

J 20½ in.

Sewage pump will mount inside here.

Toilet will mount here.

VERIFY SPECS BEFORE INSTALLING

1 Rough in the framing for the floor if desired, leaving the tower exposed in the room.

Rough in the floor framing, and build a wall to hide the tower from view. (Photo courtesy Zoeller Co.)

or

2 For added drains, drill a hole the exact size of the supplied grommet and at the exact location indicated by the manufacturer.

3 Insert the 2-in. supplied grommet and a 2-in. drainpipe for the shower, tub, or sink.

The Qwik-Jon pump-up system comes as a kit. (Photo courtesy Zoeller Co.)

CHOOSING THE TOWER LOCATION

Whether you put the tower in front of or behind a wall is a matter of aesthetics, sound, and maintenance. Locating it behind a wall will look neater, and its operation will be quieter, but you'll have to build the wall with access for maintenance of the pump and switch. You can also frame the unit and finish the entire surface, including the flat pedestal. The pedestal is 5½ in. high, just right for 2x6 joists. You can raise the entire floor or just the area around the toilet and tub/shower, if you're using the unit to support a full bath.

INSTALLING AN UP-FLUSH SYSTEM (CONTINUED)

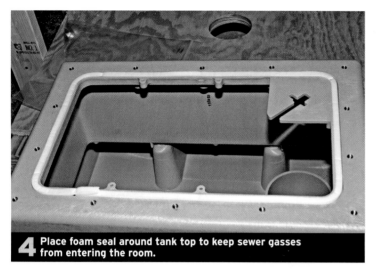

4 Place foam seal around tank top to keep sewer gasses from entering the room.

5 Attach the supplied toilet flange with wax onto the platform and build the finished floor around it.

6 Insert the supplied drainpipe with weep hole into sewage pump.

7 Insert the pump and switch into the tank.

To keep the sewer gasses inside the tank, place a foam seal around the tank top **4**, then mount the toilet flange directly to the tank, with the finished floor (if any) to be built around it **5**.

Find the pipe with the weep hole and glue it to its adaptor if it's not preassembled. Teflon-tape the threads of the adaptor and screw it into the sewage pump outlet **6**. Set the pump into the tower with the weep hole angled about 45 degrees to the left of dead-ahead center **7**.

→ **See "Aligning the Weep Hole," on the facing page.**

Then mount the switch to the right of the pump, with the switch locked in its proper position. Plug the pump cord into a 120v GFCI electrical outlet and test the system by running a temporary drain line from the pump to the outdoors, and sending water into the system with a hose **8**,**9**.

→ **See "Testing the System," on the facing page.**

Disconnect the temporary drain line and set the lid on the tank, feeding the cord through the supplied grommet. Last, connect the 2-in. drainpipe (with its check valve), and cut in a 3-in. vent line from the tank into the house system or run it outdoors. The vent pipe will fit into a grommet inserted in the large vent hole on top of the tower **10**. Do not diminish the diameter of either pipe. With all the pipes in place, lock the unit down with the manufacturer's bolts **11**.

Temporary pipe and T system running outside to test and adjust system (optional).

8 Run water to test and adjust the system.

9 Measure the water in the tank to see whether you need to adjust the pump.

10 Connect a drain (with backflow valve) and a vent line from the tank.

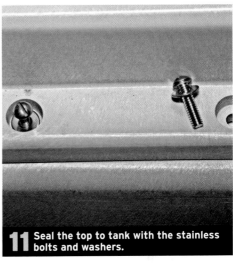

11 Seal the top to tank with the stainless bolts and washers.

ALIGNING THE WEEP HOLE

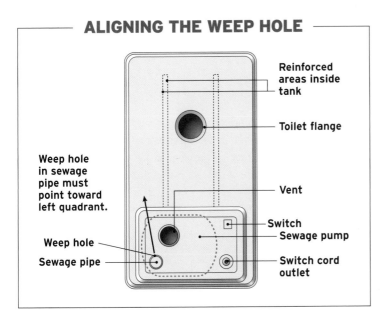

Reinforced areas inside tank

Toilet flange

Weep hole in sewage pipe must point toward left quadrant.

Vent

Weep hole

Sewage pipe

Switch
Sewage pump

Switch cord outlet

Testing the system
To make sure the system turns on and off at the water levels required by the manufacturer, adjust the pump switch by sliding the rubber grommets up and down on the rod.

Grommets

INSULATING WATER LINES

Foam insulation comes in diameters, **iron pipe size (IPS)**, and copper tubing size **(CTS)**. CTS also works for CPVC and PEX.

1 Cut the foam insulation to length using a scissors, a ratcheting scissors, or a sharp utility knife.

2 Open the insulation and slide it on one end of the pipe.

3 Push the insulation to the other end of the pipe and bring the edges together along its entire length.

4 Tape the seam or compress the edges with the self-adhesive strip.

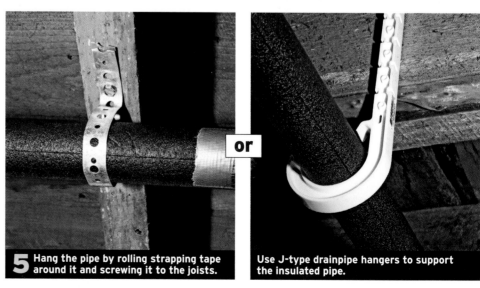

5 Hang the pipe by rolling strapping tape around it and screwing it to the joists.

or

Use J-type drainpipe hangers to support the insulated pipe.

Insulating water lines does not stop them from freezing, it just delays it. The logic of insulating water lines is based on the hope that you can out wait the cold long enough for warmer temperatures to prevail—for example, when temperatures fall below freezing at night but rise above freezing during the day when the sun comes out.

There are two materials you can use to insulate water lines. You can either wrap them with fiberglass insulation (which rarely works very well) or snap on split-foam insulation. If you're serious about insulating water lines, use the snap-on foam, the thicker the better. Some varieties come with a peel-off backing that exposes a self-adhesive strip on the edges, sticking the edges together. Absent the self-adhesive, you must use duct tape on the edges at intervals along its entire length.

To start, cut the foam insulation to length and slip it onto the pipe ❶,❷. Close the seam by peeling off the protective adhesive strip and pressing the edges together or by taping the seam ❸,❹. To hang an insulated water line, use pipe hanger strapping or plastic J-hooks ❺. To insulate an elbow, either cut a slit out of the pipe and slide it onto the corner or cut a diamond pattern out of the foam and slide the cutout over the elbow.

Turning a corner

To insulate an elbow, you can either cut out a triangular section of the insulation and slide the cutout to fit around the elbow, as shown (right) or cut a rectangle out of a section and slide it on the elbow, as shown (below). Use whatever method makes it easiest to push the insulation around the elbow.

Cut a diamond shape **in the foam about 3 in. from the elbow.**

Slide the cutout **up to and around the elbow.**

Butt the two ends together, **and note the distance to the corner.**

Cut a slot of that length **(from one pipe only).**

Rotate and slide the cutout **over the elbow. Use the cut piece to plug the end.**

OUTDOOR PLUMBING

WHEN YOU STOP TO THINK about it, a lot of plumbing is installed outdoors. Water lines go to outbuildings, the house, wells, gardens, animal water bowls, outside showers, or just a place to wash the car. And running outdoor plumbing can be easier than tackling an indoor installation—laying pipe in an open trench doesn't require working in or around walls or building them.

You can use rigid pipe, such as PVC, but PVC is hard to work with, cracks easily, and tends to leak at threaded joints under pressure. By far, the best pipe for $3/4$-in.

and 1-in. underground water lines is polyethylene. For $1^1/2$-in. lines, it is a toss up between poly and PVC, but for 2-in. water lines, PVC is a must.

Although it may seem that all you need to do is dig your trenches and lay the pipe and fittings, you'll save yourself a bunch of headaches by sketching out your plan first and making sure your pipes are below the frost line in your area and deep enough to be out of the way of your garden spade or rotary tiller.

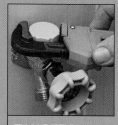

POLYETHYLENE PIPE

Polyethylene is a plastic pipe used underground to bring water to the house or to take water from the house to an outside buried hydrant or an outbuilding. It is not approved for hot water use or anywhere inside the house, except to connect to the pressure tank or to the main house cutoff valve. It is normally black in color and comes in rolls.

Grades of polyethylene

Grades (quality) of polyethylene are designated by how much pressure, in pounds per square inch (psi), the pipe is designed to take. The thicker the walls of the pipe, the larger the psi, and the higher the pressure. When you lay polyethylene in a ditch, you want thick pipe to minimize the possibility of leaks developing over time. Be sure to use 160 psi minimum whenever possible and always clear the ditch of rocks and sharp objects before you lay the pipe. Then lay in a bed of sand or install the pipe in flexible drainpipe, and when you refill the ditch, don't let rocks fall back in on the pipe.

Polyethylene fittings

Polyethylene uses barbed fittings that slip into the pipe, and the pipe is tightened on the fittings with stainless-steel clamps (use marine-grade all-stainless hose clamps). Fittings can be brass, plastic (which can bend and crack), stainless (the best, but hard to find), or galvanized steel (which tends to rust). You may have to warm the pipe to get it to slip over the barbs. Use a common hair dryer or the heat (not the flame) of a torch.

Polyethylene pipe comes in rolls of various lengths and is rated for different pressures.

Galvanized insert fittings are quite pretty when new (right), but will rust heavily (left).

This special brass union male adapter is used when there is no room to bend the pipe during installation.

WORKING WITH POLYETHYLENE PIPE

- To straighten rolled pipe, tie one end to a fixed object and unroll it. The pipe will tend to straighten itself in about 1 hour.

- Always use two clamps on the end of each fitting shank to secure the pipe to the fitting.

- In a trench, never lay polyethylene pipe on rocks or sharp objects, only on a soft bed of sand or dirt or inside a run of flexible drainpipe. To slip polyethylene pipe into a drainpipe, cut a tennis ball and place it over the end of the polyethylene pipe. This will keep the end from snagging on the drainpipe corrugations.

- Plastic male adapters tend to leak at the threads. Most pros use brass or stainless steel.

- To keep the outside surface of galvanized fittings from rusting, wrap them with electrical tape.

Polyethylene pipe grades vary with thickness. From left to right, 100 psi (1/2 in.), 160 psi (3/4 in.), and 160 psi (1 in.).

Plastic insert adapters. Avoid those with few barbs (left) and choose those with completely barbed shanks (right).

Brass insert adapters are preferred over galvanized and plastic.

Avoid nylon insert adapters. They are very soft and damage easily.

The ultimate male adapter for polyethylene pipe is made from stainless steel.

ACCORDING TO CODE

Codes will vary in the minimum psi allowed for underground pipe. Pipe rated for 100 psi is typically the minimum. But experience has shown that 160-psi pipe causes the least trouble.

INSTALLING FITTINGS

Most toothed blades, like a hacksaw blade, will leave jagged edges so it's best to cut poly pipe with a ratcheting scissors **❶**.

You'll need to soften the end of the pipe so it will slip easily on the fitting. Use either a hair dryer or the heat from a torch (not the flame) **❷**. When the pipe end is very warm, push the fitting into it. Use a rubber mallet to seat the fitting until all the barbs are inside the pipe **❸**. Slip two marine clamps over the barbs, rotating the clamp heads to opposite sides to distribute the clamp pressure evenly on the barbs. Tighten the screw with screwdriver or $5/16$-in. hex-head driver **❹**.

1 Cut the pipe to length with a ratcheting scissors.

2 Heat the cut end with a hair dryer or torch (the heat, not the flame).

3 Push the fitting fully into pipe, seating it with a rubber mallet, if necessary.

4 Tighten two marine clamps on the pipe over the fitting barbs.

STAINLESS CLAMPS

Polyethylene pipe is fastened to fittings with stainless-steel clamps. Both the band and screw must be stainless, not just the band. If only the band is stainless, it's an automobile hose clamp, not a marine clamp, and it will rust when buried in the ground. Look for "all stainless" stamped on the band. If the clamps come in a blister pack, read the instructions on the back to see if the screw is stainless.

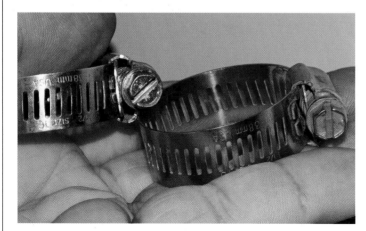
If the screw is anodized, as it is in the clamp on the left, do not use the fitting in underground plumbing.

Look for "all stainless" on the band (upper clamp). Nonstainless clamps will carry only the manufacturer's name (bottom clamp).

INSTALLING BURIED HYDRANTS

Buried hydrants (also called remote or frost-free hydrants) allow you to bring running water to locations away from the house. You can install them either at the end of an underground pipe run or in line, with multiple hydrants at different locations.

Choosing a hydrant

To keep the water from freezing, you must install the hydrant valve below the frost line. First find out the depth of the frost line from your local building department. Then purchase a hydrant specified for that depth. For example, if the frost line is down 2 ft., you want a "2-ft. bury" hydrant. If it's down 3 ft., you want a "3-ft. bury" hydrant.

How a buried hydrant works

A hydrant fitting on the bottom end will have a small drain hole and a $3/4$-in. female thread for the water supply. When the handle is turned off, the valve seals at the bottom, and the leftover water in the column drains out the drain hole so it won't freeze and split the pipe.

>> >> >>

Buried hydrants **let you bring water to locations away from the house.**

At the bottom of the hydrant, the fitting has a drain hole and threads for the water line. Use an interface appropriate to the use of the hydrant and the kind of pipe you're using.

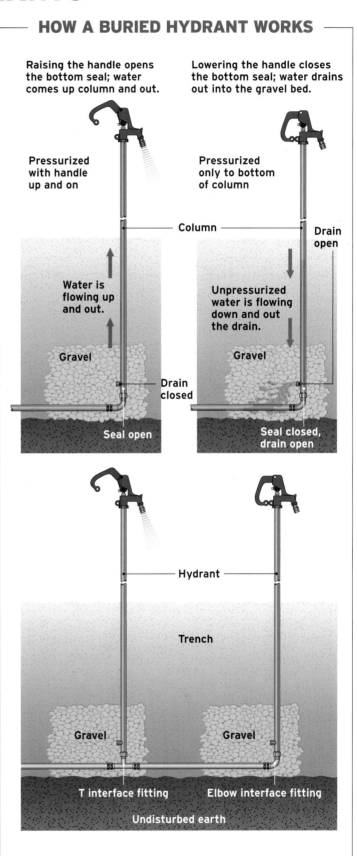

HOW A BURIED HYDRANT WORKS

Raising the handle opens the bottom seal; water comes up column and out.

Lowering the handle closes the bottom seal; water drains out into the gravel bed.

Pressurized with handle up and on

Pressurized only to bottom of column

Column

Drain open

Water is flowing up and out.

Unpressurized water is flowing down and out the drain.

Gravel

Gravel

Drain closed

Seal open

Seal closed, drain open

Hydrant

Trench

Gravel

Gravel

T interface fitting

Elbow interface fitting

Undisturbed earth

INSTALLING BURIED HYDRANTS (CONTINUED)

Always use two clamps to hold pipe onto poly fittings.

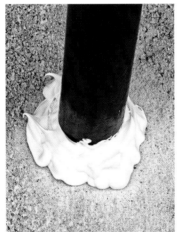

When taking the hydrant water line through a concrete wall, protect it with a PVC sleeve. Cut the hole in concrete with a rotary hammer, insert the PVC sleeve and the poly pipe, and caulk around the hole.

INSTALLING OUTDOOR PIPE

If possible, lay pipe on soft dirt without sharp rocks (fill has to be without sharp rocks as well). If not possible, lay pipe on layer of sand that is placed over the sharp rocks. If not practical, slide poly pipe in corrugated drainpipe and lay that in a ditch.

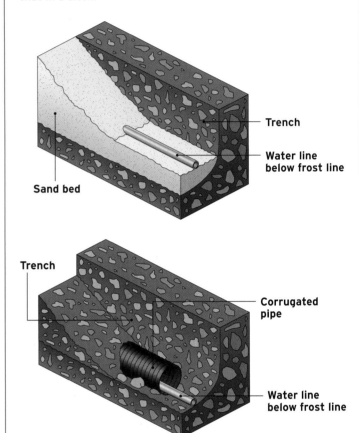

Use this method when there are too many rocks to remove.

Laying pipe in a ditch

Dig a trench from the house to the location(s) of the hydrant(s). Make sure to dig slightly below the frost line. You can excavate short runs with a shovel. For long runs, to save time and your back, rent a trencher. To keep water line movement against rocks from wearing holes in the pipe, lay in a 2-in. to 3-in. bed of sand or install the poly water line in a flexible drainpipe. Once you've installed each hydrant, shovel gravel around the base and cover the pipe with another layer of sand. Then refill the trench and replace any sod.

Installing hydrants

Installation for both in-line (using T-fittings) and end-of-the-run hydrants (using elbows) is basically the same. Tape the interface fitting threads with Teflon tape, screw the fitting into the hydrant, slip the pipe on the fitting, and secure it with two all stainless clamps on each side. Set the hydrant in the trench, drop a bucket of gravel around drain hole, and fill in the ditch. To keep an end-of-the-run hydrant turned in the right direction, make the turn with the fitting, not the pipe.

Taking a hydrant water line through a wall

When taking the lines to the hydrant through a concrete or block basement wall, punch out a hole for a PVC pipe slightly larger than the water line pipe. Use the PVC as a sleeve through the wall to protect the water line pipe from abrasion. Once in, caulk all around the joint to prevent water intrusion.

Hydrant interfaces

How you interface your hydrant will depend more on where it fits in the outdoor line than on the pipe material (assuming it's poly.) Use T-fittings on in-line runs **A,B,C** and elbows on end-of-the-run installations **D,E.** Buy and assemble the best fittings you can afford. An investment now will save maintenance, costs and labor later.

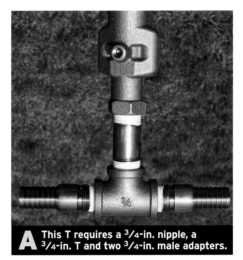

A This T requires a ³/₄-in. nipple, a ³/₄-in. T and two ³/₄-in. male adapters.

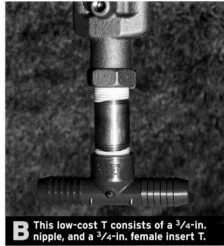

B This low-cost T consists of a ³/₄-in. nipple, and a ³/₄-in. female insert T.

C The best T, an all-in-one ³/₄-in. fitting. Male threads and male inserts for the T.

D This end-of-run assembly consists of a street elbow and a male insert adapter.

E The best end-of-the line fitting is a ³/₄-in. elbow, male threads to male adapter.

Repairing a leaky hydrant

If the buried hydrant is dripping like a faucet, its seal needs to be replaced. This is not easy because the factory-installed head is extremely tight, and it may not be possible to remove the head to get to the seal. To stop a dripping hydrant, just screw a hose bibb cap on the threads.

A hydrant may leak around the packing nut as soon as you start using it. To stop the leak, tighten the nut with a wrench.

INSTALLING A WALL-MOUNTED FROST-FREE FAUCET

Through-the-wall hydrants are installed on almost every house. They are good for washing the car, watering the flowers, and other outdoor chores. With frost-free hydrants, the handle is on the outside of the house and the seat is on the inside. When you turn the hydrant off, the water in the column drains forward and out, preventing the hydrant from freezing and bursting. You must remove the hose from the hydrant in the winter, however. If you leave a hose attached, water from the hydrant will drain into the hose, and a hard freeze may burst it.

To install a wall-mounted frost-free unit, you need to first attach the interface fitting for the kind of pipe in your house (copper, PEX, or CPVC). The easiest way is to screw on a universal fitting (such as a Sharkbite) that fits all **❶**. However, most hydrants, even though they have ½-in. male threads, will also accept copper sweat-fittings on the inside, if you prefer.

➡ See "Push-on Fittings," pp. 40–41.

Once you've attached the fitting, measure its outside width **❷**, and using a rotary hammer (in masonry) or a hole saw (in the rim joist), open a hole in the wall slightly larger than the fitting **❸**. Typically you want the hole to be about the height of a 5-gal. bucket off the ground. Dry-fit the hydrant to make sure it will slip all the way into the hole **❹**. Level the hydrant, and mark the screw holes on each side. Withdraw the hydrant, and drill the screw holes **❺**. Put some silicone sealant around the hole on the wall, and insert the hydrant **❻**. Remove the handle and drive in the attachment screws **❼**. Reattach the handle **❽**, and connect the hydrant's pipe into the house pipe, and you are done **❾**.

1 Pick out your interface fitting, tape the hydrant's threads, and attach the fitting to the hydrant.

4 Dry-fit the hydrant in the hole to make sure it will fit without binding.

7 Remove the handle and install the attachment screws.

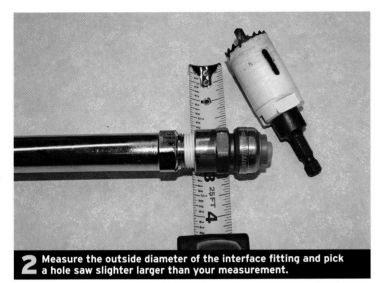

2 Measure the outside diameter of the interface fitting and pick a hole saw slighter larger than your measurement.

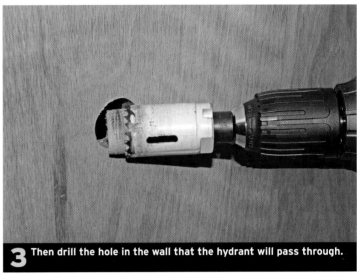

3 Then drill the hole in the wall that the hydrant will pass through.

5 Level the hydrant body mounting holes and mark their location. Withdraw the hydrant and drill the screw holes.

6 Put some sealant (rated for your climate) around the drilled hole and insert the hydrant.

8 Reattach the handle.

9 Attach whatever pipe you want for house pipe to the hydrant and you're done.

PROTECTING OUTDOOR PLUMBING FROM FREEZING

1 Slip the loop of the string over the handle of the faucet.

2 Push the box against the wall over the faucet and slide the lock mechanism up the string to fit snug against the box.

A Heat tape is designed to keep water lines from freezing.

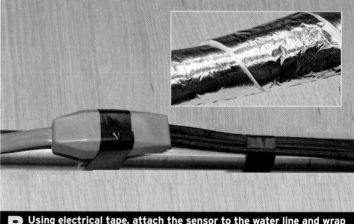

B Using electrical tape, attach the sensor to the water line and wrap tape at 1-ft. intervals. Then wind fiberglass insulation on the pipe.

For those who don't have a frost-free outside faucet, a frost cap will keep the faucet from freezing, and heat tape will protect water lines in winter temperatures.

Installing a frost cap

To install a frost cap, slip the oval end of the string inside the cap over the faucet handle **1**. Place the box over the faucet and pull the string out the front of the cap. Keeping the cap tight against the wall, slide the plastic lock up against the front of the box **2**.

Installing heat tape

In the past, the use of heat tape was controversial because it caused fires. Today's low-wattage heat tape is less susceptible to overheating **A**, but you still have to use care to not wrap it on itself, and you must keep its sensor tight against the pipe.

A space-age product called Frostex® used every inch of its cable as a heat sensor, and you could wrap it over itself without fear of fire. It was available for many years at local hardware stores and was sold by the foot. But price increases have forced it into the large internet wholesaler market, now available only in 100-ft. rolls. Though it is arguably the best designed heat tape, few people will buy a roll just to get 10 ft.

To run a single-sensor heat tape, start by taping the sensor on the pipe **B**. Check manufacturer's instructions. If allowed, tape the heat tape to the pipe at 12-in. intervals and wrap fiberglass insulation over the entire length.

FIXING A LEAKY VACUUM BREAKER

A vacuum breaker on an outside hose bibb is required in most areas to protect your in-house water supply from contamination. Such contamination can occur, for example, when spraying liquid fertilizer on the lawn—if the water pressure in the house dropped, for example, the fertilizer could flow backward into your water lines.

To repair a leaky breaker, remove the plastic cap with a small pipe wrench or large pliers ❶. Unscrew the slotted white plastic disk with the tips of needle-nose pliers ❷. Remove and clean the O-ring, and replace it if it is nicked or cut ❸. Clean the center seal, and replace it if it's damaged ❹.

1 Using a pipe wrench or large pliers, turn the cap counterclockwise and remove it.

2 Using needle-nose pliers, remove the white disk.

3 Check the O-ring in the plastic disk for cuts.

4 Using needle-nose pliers, remove the small center seal and clean it.

ACCORDING TO CODE

All codes require a vacuum breaker on an outside hydrant.

SCREW-ON VACUUM BREAKERS

Screw-on vacuum breakers are made for hose bibbs without vacuum-breaker or antibackflow devices. To make your hose bibb antibackflow, simply screw on a breaker (below left) and tighten the tab to lock it (below right). The tab will break away.

CONTROLLING WATER-SYSTEM PRESSURE AND FLOW

Too much water pressure coming into your house can cause the T&P valve on your water heater to open. Municipal pressures vary but your house plumbing can take no more than 80 lb./psi, and pressures above that (caused by being too close to a tower or a pump) require a pressure regulator.

To check the pressure, screw a water-pressure gauge on a hose bibb. Turn the center brass knob to set the red needle to 0 psi ❶. Turn the hose bibb on and wait until the black needle settles at its highest mark. As it moves, it will take the red needle with it ❷. Turn off the faucet, and remove the gauge (so prolonged exposure to sunlight won't damage it) ❸. The red needle will

indicate the maximum water pressure. If the pressure is above 80 psi, you'll need to cut a regulator directly into the main water line, either outside or just inside the house. In either location, use the same procedures you would to cut in a T-fitting, and place the regulator where you can adjust and maintain it.

➜ **See "Installing T-Fittings," p. 52.**

Adjusting the regulator

Once installed, the regulator will need to be adjusted to the pressure you want in the house. First loosen the lock nut on top of the dome. Then adjust the bolt on top of the threaded shaft to obtain the desired pressure.

1 Set the red needle to zero psi by turning the center brass knob counterclockwise.

2 Turn the faucet on. The black and red needles will move simultaneously.

3 Turn the faucet off and remove the gauge.

Incoming excessive city water pressure 90 psi

Reduced or regulated water pressure 60 psi

Water flow

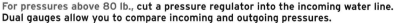
¾-in. T, ¾-in. by ¼-in. bushing, 100-lb. gauges

For pressures above 80 lb., cut a pressure regulator into the incoming water line. Dual gauges allow you to compare incoming and outgoing pressures.

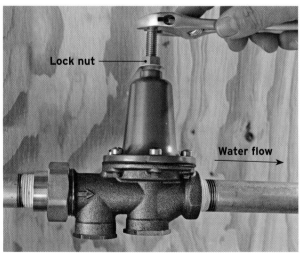

Lock nut

Water flow

Loosen the lock nut and adjust the pressure by turning the bolt on top the dome. Retighten the lock nut.

Check valves

The purpose of a check valve is to stop the reverse flow of water, and the most common valve has a single seat and spring **A**. A dual check valve has two of the same and is required if you are using it to protect city or well water from possible contamination from your house **B**. In the country, without a check valve, you could contaminate your well and the entire aquifer. On a city water supply, you could contaminate the water of your entire block. For example, suppose you're mixing bug killer with a hose submerged in a drum of water. If a water main broke on the street downhill from you, all the bug killer mix in the drum would flow backward, through your house and out into the main water lines in the street.

A check valve can be buried, as long as it is not made with a drain. Some areas don't require antibackflow devices in the main line, as long as each hose bibb has a vacuum breaker, which has the same backflow prevention function.

A A common spring-loaded check valve. It can be installed in any position, but water flow must be in the direction of the arrow.

B A dual check valve is required by code in most areas as protection against accidental contamination of a public water source.

Controlling dual water sources
Some installations require a dual water source (drawing on left) or a single source with two users (drawing on right). To allow only one water source into the house at a time (drawing on left), use two ball valves and a brass tank T. Screw a ball valve in the two inputs of the T (use any two ends), and interface the two water sources into the ball valves. Take the water to the house via the one open T end. You must have only one valve open at a time—never close both valves unless you intend to cut off all water flow. To keep from accidentally backfeeding one source into the other if accidentally opening both valves, attach check valves to the ball valves.

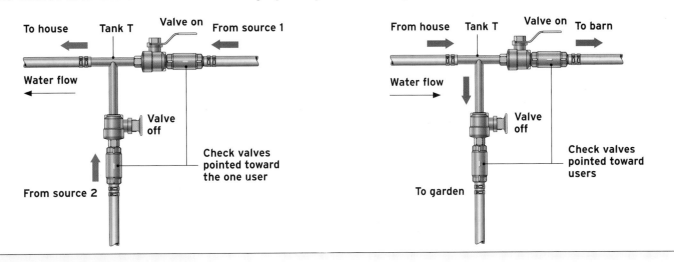

GLOSSARY

ABS An early plastic pipe material used as drain pipe, ABS (acrylonitrile-butadiene-styrene) has now been superseded in many municipalities in favor of PVC drainpipe. Check your local codes.

ADAPTER A fitting that makes it possible to transition from male to female endings, or vice-versa, allowing joints between different kinds of pipe.

BALL VALVE Commonly used as a full-flow main turn-off valve. Named after a ball on the inside of the valve that controls the water.

BUFFALO BOX A type of whole-house shutoff in which the valve is in a plastic or concrete box set in the ground.

CHECK VALVE A one-way water flow device. Water is allowed to flow in one direction only.

CLEANOUT A fitting in a trap or drainpipe that allows access for clearing blockages.

CLOSET BEND The elbow-fitting beneath a toilet that carries waste to the main drain.

COUPLING A fitting used to connect two lengths of pipe in a straight run.

CPVC Chlorinated polyvinyl chloride pipe, approved by many municipalities for indoor hot and cold supply lines.

DIELECTRIC FITTING A fitting for joining dissimilar metals. Its plastic washer insulates the pipes from each other, preventing a corrosive chemical reaction.

DIVERTER A faucet valve that diverts water from a faucet spout to a hand sprayer or from a tub spout to a shower head.

DUAL CHECK VALVE Akin to a check valve but has two one-way spring-loaded check valves inside. Required by most codes to isolate the house water system from the public water system.

ELBOW A 90-degree fitting that changes the direction of a water supply line.

FALL The slope of a drain line, generally a minimum of 1/4 in. per foot.

FEMALE FITTING A fitting that has threads on the inside.

FITTING Any connector (except a valve) that joins pipes.

FIXTURE Any device that provides water or sanitary disposal of wastes.

FIXTURE DRAIN The drainpipe and trap leading from a fixture to the main drain.

FLOW RESTRICTOR A device that restricts the flow of water and reduces water use.

FLUSH VALVE The assembly inside a toilet tank that releases water into the toilet bowl to start the flushing action.

FLUX A chemical paste applied to copper and brass pipes and fittings that improves the cleaning and bonding processes for soldering.

GATE VALVE Commonly used as a full-flow main turn-off valve. Named after its rising and falling gate that controls the water.

HOSE BIBB A threaded faucet to which a hose can be attached.

ID Abbreviation for inside diameter. See also OD.

KNOCKOUTS Partially incised holes in a sink flange that allow removal for the addition of a fixture such as a sprayer.

LOOP VENT A vent in a kitchen island that loops under the island and connects to a stack from an under-floor vent line.

MAIN DRAIN The drainpipe between the fixture drains and the sewer drain.

MALE FITTING A fitting with threads on the outside.

NIPPLE A 12-in. or shorter pipe threaded on both ends and used to join fittings.

NO-HUB FITTING A neoprene gasket with a stainless-steel band that joins PVC drain pipe to ABS or cast-iron pipe.

OD Abbreviation for outside diameter. See also ID.

O-RING A round rubber ring that creates a watertight seal, chiefly around valve stems.

PB POLYBUTYLENE PIPE, A plastic pipe no longer code in the United States.

PE Flexible polyethylene supply pipe, often used in underground outdoor supply systems.

PEX Cross-linked polyethylene plastic tubing used for radiant floor heating and hot and cold supply lines.

PACKING A plastic or metallic corded material used chiefly to seal faucet stems.

PIPE-THREAD TAPE A Teflon pipe-thread wrapping that seals a threaded joint.

PLUMBER'S PUTTY A paste material used as a sealer.

PRESSURE REDUCER VALVE Required by public water systems to limit the incoming house pressure to 80 psi or less.

PSI Abbreviation for pounds per square inch; a measure of pressure levels.

PVC Polyvinyl chloride pipe, a commonly accepted plastic drainpipe, also sometimes used for outside water lines.

REDUCER A fitting sized differently at each end to join larger to smaller pipes.

REVENT A pipe that connects a fixture drainpipe to a main or secondary vent stack.

RISER A supply pipe running to an overhead fixture.

ROUGH-IN The first stage of a plumbing job, during which supply and drain-waste-vent lines are run to their destinations.

RUN Any length of pipe or pipes.

SANITARY FITTING Fittings used to join drain-waste-vent lines.

SEAT GRINDER A tool made specifically for smoothing the seat of a faucet valve.

SELF-RIMMING SINK A drop-in kitchen or bathroom sink made with a formed lip that rests on the countertop to support the sink.

SOIL STACK A vertical pipe that carries waste to the sewer drain. The main soil stack is the largest vertical drain line of a building.

STAND PIPE A 2-in. vertical pipe that connects a washer drain hose to the drain system.

STOP VALVE A valve installed in a fixture supply line that allows shutting off the water when making repairs.

TAILPIECE That part of a drain that runs from the fixture drain outlet to the trap.

T A T-shaped fitting cut into a line to change the direction of the water by a 90-degree angle.

TRANSITION FITTING A fitting that joins pipe of dissimilar materials, such as copper and plastic, and cast iron or galvanized steel and copper.

TRAP Part of a fixture drain that prevents sewer gases from entering a home interior.

TY FITTING A combination T and Y fitting. Allowed where T fittings are not.

UNION A fitting that allows easy disassembly of pipe.

VACUUM BREAKER Installed on some spigots to prevent accidental siphoning of water into the house supply lines.

VENT Vertical portion of a drain line that permits sewer gases to rise out of the house.

VENT STACK The uppermost portion of a vertical drain line through which gases pass to the outside.

WATER HAMMER A noise caused by sudden changes in the flow of water, causing pipes to repeatedly "hammer" against a framing member or against themselves.

WATER HAMMER ARRESTER A shock absorber that prevents water hammer.

WET WALL The main wall of a room that contains supply and drain lines.

Y A Y-shaped fitting.

INDEX